The Environmental Crunch in Africa

Jon Abbink
Editor

The Environmental Crunch in Africa

Growth Narratives vs. Local Realities

Editor
Jon Abbink
African Studies Centre
Leiden University
Leiden, The Netherlands

ISBN 978-3-319-77130-4   ISBN 978-3-319-77131-1 (eBook)
https://doi.org/10.1007/978-3-319-77131-1

Library of Congress Control Number: 2018934660

© The Editor(s) (if applicable) and The Author(s) 2018
This work is subject to copyright. All rights are solely and exclusively licensed by the Publisher, whether the whole or part of the material is concerned, specifically the rights of translation, reprinting, reuse of illustrations, recitation, broadcasting, reproduction on microfilms or in any other physical way, and transmission or information storage and retrieval, electronic adaptation, computer software, or by similar or dissimilar methodology now known or hereafter developed.
The use of general descriptive names, registered names, trademarks, service marks, etc. in this publication does not imply, even in the absence of a specific statement, that such names are exempt from the relevant protective laws and regulations and therefore free for general use.
The publisher, the authors and the editors are safe to assume that the advice and information in this book are believed to be true and accurate at the date of publication. Neither the publisher nor the authors or the editors give a warranty, express or implied, with respect to the material contained herein or for any errors or omissions that may have been made. The publisher remains neutral with regard to jurisdictional claims in published maps and institutional affiliations.

Cover credit: Mike Korostelev www.mkorostelev.com/gettyimages

Printed on acid-free paper

This Palgrave Macmillan imprint is published by the registered company Springer International Publishing AG part of Springer Nature
The registered company address is: Gewerbestrasse 11, 6330 Cham, Switzerland

Acknowledgements

This book has its origins in a very lively and successful panel at the biannual European Conference on African Studies (ECAS), held in Paris, France, in July 2015. The final text, however, is quite removed from this event, with new authors and chapters included.

As editor, I wish to thank the contributors to this book for their patience and cooperation. I also express my profound gratitude to Palgrave Macmillan, in particular Ms. Rachael Ballard and Ms. Joanna O'Neill, for their good advice and expert supervision of the publication process. I am also grateful to the reviewers of the book project proposal for their critical points and advice.

Contents

1   Introduction: Promise and Peril in Africa—Growth Narratives vs. Local Environmental Problems   1
Jon Abbink

2   Cash for Cashews: Does It Add Up?   29
Margaret L. Buckner

3   Trade-Offs Between Crop Production and Other Benefits Derived from Wetland Areas: Short-Term Gain Versus Long-Term Livelihood Options in Ombeyi Watershed, Kenya   51
Serena A. A. Nasongo, Charlotte de Fraiture and J. B. Okeyo-Owuor

4   Agriculture, Ecology and Economic Development in Sub-Saharan Africa: Trajectories of Labour-Saving Technologies in Rural Benin   85
Georges Djohy, Honorat Edja and Ann Waters-Bayer

5 Is Growing Urban-Based Ecotourism Good News
   for the Rural Poor and Biodiversity Conservation?
   A Case Study of Mikumi, Tanzania   113
   Stig Jensen

6 Losing the Plot: Environmental Problems and
   Livelihood Strife in Developing Rural Ethiopia—Suri
   Agropastoralism Vs. State Resource Use   137
   Jon Abbink

7 Cameroon's Western Region: Environmental Disaster
   in the Making?   179
   Moses K. Tesi

8 The Impasse of Contemporary Agro-pastoralism
   in Central Tanzania: Environmental Pressures
   in the Face of Land Scarcity and Commercial
   Agricultural Investment   207
   Tadasu Tsuruta

9 Down by the Riverside: Cyclone-Driven Floods
   and the Expansion of Swidden Agriculture
   in South-western Madagascar   241
   Jorge C. Llopis

10 Challenging Impediments to Climate Change
   Initiatives in Greg Mbajiorgu's *Wake Up Everyone*   269
   Norbert Oyibo Eze

11 Future in Culture: Globalizing Environments
   in the Lowlands of Southern Ethiopia   287
   Echi Christina Gabbert

Index   319

Notes on Contributors

Jon Abbink is senior researcher at the African Studies Centre Leiden and professor of Politics and Governance in Africa at Leiden University, the Netherlands. Previously, he was research professor in African Studies at the VU University, Amsterdam. His anthropological-historical research has primarily been carried out Northeast Africa, especially Ethiopia. His main interests are political anthropology, developmentalism in Africa, the ethnology and cultural history of (Southwest) Ethiopia, and religious culture and the public sphere in the Horn of Africa. His most recent books are the co-authored *Suri Orature* (2013) and *A Decade of Ethiopia: Politics, Economy and Society 2004–2016* (2017).

Margaret L. Buckner recently retired after twenty years as Professor of Anthropology at Missouri State University. She earned her masters and Ph.D. in Ethnology at the Université de Paris X-Nanterre, where she is still a member of the Laboratoire d'ethnologie et de sociologie comparative (UMR 7186). Her graduate studies focused on the Zande of Central Africa, and she recently published two books on Zande oral tradition: *Poètes nzakara*, vol. 2 (2015), and *Touré le Farceur. Chantefables zandé* (2017). In the early 1990s, she worked as a medical anthropologist in Guinea Bissau and has been carrying out ethnographic fieldwork there ever since.

Charlotte de Fraiture holds a Ph.D. in Civil Engineering (specialisation Water Resources Management) from the University of Colorado in Boulder, USA. Her M.A./M.Sc. studies were in economics and in irrigation water engineering. She has over 20 years of international working experience in the field of water management for agriculture. She was the project leader of several research projects, among others a large BMGF funded project on identifying promising options for smallholder water management in Ghana, Burkina Faso, Tanzania, Ethiopia, Zambia and India. Besides research she fulfilled several management tasks, being Head of Office of the West Africa office, Theme Leader and Group Head. Since January 2012, she joined IHE Delft as professor of Land and Water Development.

Georges Djohy is agricultural engineer and anthropologist. He studied Agricultural Economics and Rural Sociology at the University of Parakou (Benin) and holds a Ph.D. in Anthropology from the University of Göttingen (Germany). He has been involved since 2008 in among Fulani pastoralists, with a focus on environmental and socio-technological changes. He has an interest in land use, climate change, local innovation and grassroots organisation politics. After defending his doctoral thesis, he served at the Göttingen Institute for Social and Cultural Anthropology (GISCA) as a Research Associate. He is currently working as a Postdoctoral Researcher at the Faculty of Agricultural Sciences, University of Parakou, Benin, studying livelihoods and structural changes, gender and power relations through grassroots associations of Fulani pastoralist women.

Honorat Edja is anthropologist and Associate Professor at the Faculty of Agricultural Sciences, University of Parakou (Benin), and researcher at the LARES. He is currently the Dean of the Faculty and has research interest in land-tenure systems, rural development and environmental change. He has contributed to many books on land-tenure systems in Benin, highlighting forms of tenure negotiation, renegotiation and security in time and space. He is an expert on development and implementation of rural land mapping, focusing on the dynamics of knowledge production, governance and participation.

Norbert Oyibo Eze is a senior lecturer and the immediate past Head of Department of Theatre and Film Studies, University of Nigeria, Nsukka. He holds a Ph.D. in Dramatic Theory and Criticism and has written three books (two for the National Open University of Nigeria) and one play. His major areas of specialisation are theatre history, dramatic theory and criticism, and dramatic literature. He has published in local and international journals and has attended and presented papers in many local and international conferences and is also an editor and a poet. On 29 July 2016, the University of Nigeria awarded him the Head of Department of the Year title for his administrative performance.

Echi Christina Gabbert is anthropologist and senior lecturer at the Institute for Social and Cultural Anthropology at Göttingen University. Her long-term fieldwork among agro-pastoralists in southern Ethiopia resulted in the award-winning Ph.D. thesis 'Deciding Peace' at the Max Planck Institute for Social Anthropology, Halle-Saale. Her research foci reach from material culture, music, oral history, interethnic relations, agro-pastoral subsistence systems and ecology, to peace and conflict studies. She currently develops the 'Global Neighbourhood approach', looking at local responses to global investments, and coordinates the 'Lands of the Future Initiative' that brings together scholars, pastoralists and practitioners from all disciplines—to raise levels of pertinent knowledge about agro-pastoralism within changing economies.

Stig Jensen is associate professor and former director at the Centre of African Studies (CAS), University of Copenhagen. He holds a M.A. in Political Science from Copenhagen University and a Ph.D. in Geography from Department of Social and Spatial Changes, Roskilde University. Stig has comprehensive experience with research, consultancy and advisor role in different setting as well as extensive fieldwork experience from Africa, Asia and South America. His most recent book is the co-authored *Higher Education and Capacity Building* *in Africa—The Geography and Power of Knowledge Under Changing Conditions* (London: Routledge, 2016).

Jorge C. Llopis is a Ph.D. candidate in Geography and Sustainable Development at the Centre for Development and Environment, at the University of Bern in Switzerland. He holds a M.A. in African Studies from the University of Copenhagen in Denmark and has a background in Geography and History. His research has mostly focused on exploring social-ecological dynamics in and around protected areas in diverse regions of Madagascar. He is particularly interested in analysing the effect of extreme weather events on land use change dynamics.

Serena A. A. Nasongo is a Ph.D. student at the University of Amsterdam, the Netherlands, in the Dutch-funded 'Ecology of Livelihoods in African Wetlands (ECOLIVE)' project. She holds an M.Phil. in Environmental Human Ecology from Moi University, Kenya, B.Sc. in Agriculture and Home Economics from Egerton University, Kenya. She has over 15 years working experience in the field of environment studies, most recently in a JICA-funded Community-based Flood Mitigation projects in Lake Victoria Basin and 'No Sex For Fish (NSFF)' Project in Lake Victoria Basin. She has been involved in numerous interdisciplinary, multinational and multicultural environmental management consultancies in Kenya.

J. B. Okeyo-Owuor holds a Ph.D. in Zool/Population Ecology/IPM, an M.Sc. in Entomology and a B.Sc. in Agriculture. He is the Founder and Director General of Victoria Institute for Research on Environment and Development (VIRED) International based in Kisumu, Kenya. He is a Professor at Rongo University College, Kenya, lecturing in environmental biology and health subjects including ecology, biodiversity and wetland management. He researches environmental and natural resources management in tropical Africa. He has over 25 years professional expertise and working experience in agriculture, ecosystem and natural resources management, food security and nutrition in Africa and was involved in the management of several international donor-funded, collaborative research and community-based projects. In 2010, he received the Presidential Award of Elder of Burning Spear (EBS) in

Kenya for his scientific and development records in the country and beyond.

Moses K. Tesi is a Professor in the Department of Political Science and International Relations at Middle Tennessee State University, Murfreesboro, TN, USA. His research focus is on the intersection between foreign policy and domestic politics in the relations between major and minor powers in international affairs; environment and sustainable development in Africa; and the politics and economics of state–society relations in African governance. He also edits the *Journal of African Policy Studies*. His most recent book is *Balancing 'Sovereignty and Development' in International Affairs: Cameroon's Post Independence Relations with France, Africa, and the World* (Lanham 2017). Among his various publications on development and environmental issues are: *Global Warming and Health: The Issue of Malaria in Eastern Africa's Highlands* (CIGI, 2011) and the edited *The Environment and Development in Africa* (2000).

Tadasu Tsuruta is Professor, Faculty of Agriculture, Kindai University, Japan. He has been working on comparative studies on rural societies in Africa and Southeast Asia. One of his specific research themes is the moral economy of rural dwellers in Africa and Asia. His publications on this theme include the co-authored *Contemporary Perspectives on African Moral Economy* (2008) and *Comparative Perspectives on Moral Economy: African and Southeast Asia* (2011). One of his recent research interests is changing agro-pastoralism in semi-arid Africa, with fieldwork projects in Tanzania. He also co-authored *Fair Trade and Organic Initiatives in Asian Agriculture* (with Rie Makita, 2017).

Ann Waters-Bayer is an international expert affiliated with PROLINNOVA (Promoting Local Innovation) International Support Team based at the Royal Tropical Institute (KIT) in the Netherlands. She holds a Ph.D. in Agricultural Sociology from the University of Hohenheim (Germany) and has worked for over 35 years with pastoralist and farmer communities in Africa and worldwide on several topics, including food and nutrition security, animal husbandry, gender, agricultural development, rural development networking and information and knowledge management. She currently works on methodologies of multi-stakeholder partnership in agricultural research and development, agricultural innovation systems, pastoralist development, ecologically oriented agriculture and natural resource management, and gender issues. She has published extensively on these topics, having (co)authored several journal articles, book chapters and edited volumes.

List of Figures

Fig. 2.1	Production of cashews (with shells) in Guinea-Bissau in tons (*Source* FAO)	31
Fig. 2.2	Men building new dike	33
Fig. 2.3	Rice granaries	33
Fig. 2.4	Cashew apples	36
Fig. 2.5	Women juicing cashew apples	38
Fig. 3.1	Map showing the location of the Ombeyi wetland, Kenya (Coordinates: 0°11'–0°19'S, 34°47'–34°57'E) (*Source* van Dam et al. (2013). Linking hydrology, ecosystem function, and livelihood outcomes in African papyrus wetlands using a Bayesian network model. *Wetlands* 33)	56
Fig. 3.2	General approach of stakeholder analysis (Adapted from Darradi et al. 2006)	58
Fig. 3.3	Percentage of respondents by sub-location	58
Fig. 3.4	Level of formal education of respondents by age group	59
Fig. 3.5	Venn diagram for Kore sub-location (*Key* white—community groups; black—CSOs; white with broken edge—government departments; gray with broken edge—faith-based organizations; gray—financial organizations; and striped black—donor)	60

List of Figures

Fig. 3.6	Mean monthly rainfall (mm) for Ahero station from 1950 to 2010	64
Fig. 3.7	Historical trends for an average production of the 5 crops in the Ombeyi wetland	69
Fig. 3.8	Percentage of respondents harvesting plant wetland resources by sub-location	70
Fig. 3.9	Percentage of respondents harvesting fish wetland resources by sub-location	71
Photo 4.1	A farmer benefiting from a tractor-ploughing service in Borodarou (Gogounou)	97
Photo 4.2	A farmer spraying land to be ploughed by draught oxen in Bagou (Gogounou)	100
Fig. 4.1	Crop areas, production volumes and yields in Gogounou from 1996 to 2016 (*Source* Designed from the database of Gogounou Extension Service 2017)	102
Fig. 4.2	Land units and land-cover change in Gogounou (1982–2012) (*Source* Designed from satellite images provided by CENATEL-Benin 2014)	103
Fig. 6.1	Lush traditional sorghum and maize fields of Suri near a village, 1992. A sight no longer seen in 2017. Note trees left standing across and near the fields	146
Fig. 6.2	Plan of the sugar plantation surfaces in the Lower Omo basin (*Source* https://www.survivalinternational.nl/nieuws/7865, from an official Ethiopian project document, 2011)	153
Fig. 7.1	Map of Cameroon's Administrative Regions	202
Fig. 8.1	Study area	209
Fig. 8.2	Flows of migration and investment in Itiso	231

List of Tables

Table 2.1	Rice vs. cashews	40
Table 3.1	Important and potentially interested parties in the Ombeyi wetland	62
Table 3.2	Wealth categories in the Ombeyi wetland and water use	63
Table 3.3	Rice cropping calendar	65
Table 3.4	Income and cost of production from selected crops in the Ombeyi wetland	66
Table 3.5	Model summary	72
Table 3.6	ANOVA	72
Table 3.7	Coefficients	74
Table 3.8	Number of respondents doing irrigation and collecting wetland products	76
Table 3.9	Schedule of meetings held with stakeholders	80
Table 4.1	Inventory of farm equipment in Gogounou	96
Table 4.2	Inventory of chemicals for cotton production from 2013 to 2017	98
Table 7.1	Cameroon's population by region, 2015	183
Table 7.2	Distances from Bafoussam capital of the Western Region to major towns in the region in kms	192
Table 7.3	Cameroon's protected areas	192

Table 8.1	Population increase in the study area	213
Table 8.2	Agricultural production and other income-generating activities of sample farmers in M Village (22 household [HH])	218
Table 8.3	Livestock keeping in Hamlet H as of August 2013	222
Table 8.4	Chronological list of land purchase by JM	226

1

Introduction: Promise and Peril in Africa—Growth Narratives vs. Local Environmental Problems

Jon Abbink

> *We are jeopardizing our future by not reining in our intense but geographically and demographically uneven material consumption and by not perceiving continued rapid population growth as a primary driver behind many ecological and even societal threats.*
>
> W. Ripple et al. (2017)

Environmental Crisis Narratives and Crisis Phenomena

The above citation from William Ripple and colleagues encapsulates contemporary feelings of malaise and ambivalence about economic growth versus global environment concerns. On the one hand—especially for Africa—there are great stories of Gross Domestic Product (GDP) growth, expansion and production increase in agriculture,

---

J. Abbink (✉)
African Studies Centre, University of Leiden, Leiden, The Netherlands
e-mail: abbink@ascleiden.nl

large-scale land acquisitions, and investments for export production, and Africa as a huge 'emerging market.' Cities are growing, more people than ever catch up on education and technology, ICT innovation is proceeding, and relatively less people stay below the poverty line. On the other hand, there are the persisting stories on the African—and global—natural environment that continues to deteriorate and impoverish, especially for local economies and environments. This book is about that tension and offers some reflections and case studies on the observed dilemmas. It is not an exhaustive treatment of the subject, which is dealt with in a now substantial literature, but contributes to the debate. There is not only incontrovertible climate change and global warming,[1] but also ongoing resource squeeze, decline of forests and biodiversity, predation and soil degradation, all ticking on steadily. Many developing countries claim the right to grow and pollute first, and only address environmental problems later—like, as they often say, China did. And if investments are to be made in green economies and sustainability, then it stands to reason the developed world should pay more. This sounds reasonable, and agreements pertaining to this have been made in the framework of the COP conferences to counter global warming, notably the 2015 COP 21 in Paris. But environmental decline—next to global warming—marches on, fueled by strong population growth in Africa and insensitive or out-of-context plans to 'modernize' agriculture and eliminate smallholder economies, pastoral economies, etc.—phenomena for which African governments should also develop policy.

This 'message' of looming problems is not necessarily a return to stories like Robert Kaplan's on 'the coming anarchy' (Kaplan 1994); a close reader of that controversial essay might note how much more nuanced reality is 25 years later. His argument was about politics rather than environmental decline as such, but he linked the two. The casual model was flawed, and he did not foresee the surge in food production (e.g., via genetically modified crops and expansion of cultivation areas), the rise of Internet as a driver of the economy and a space of critique (start-ups, spread of market information, and social media), and green technology. But Kaplan was right about the fact that many areas became harder and harder to govern, and that 'civil society' in the developing world remained very precarious and weak—and this included organizations and powers in the domain of environmental governance. And even with those new technological means and human inventions, land surfaces today keep filling up,[2] freshwater reserves keep declining, oceans are more polluted, species are disappearing, and resource struggle continues. There are areas that recover or retain vegetation (see Tiffen et al.'s classic study 1994; Brandt et al. 2017), but there *is* no infinity of resources. Biophysical ecologies are gradually declining: in relative terms, but most probably also in absolute terms.

With an *annual* population growth in Africa of more than 31 million (and rising),[3] it is not difficult to see that the cliché of African environments 'facing major challenges,' and in many places outright destruction, is true. The environmental and social-scientific literature on it is accumulating.[4] While deterministic and overly alarmist accounts have been nuanced by many studies of situated recovery and resilience of both environmental systems and affected people (cf. Fairhead and Leach 1996; Burger and Zaal 2009; Abbink 2017, and many others), problems of environmental sustainability (i.e., long-term reproducibility of living conditions) and management have not gone away; indeed they are now omnipresent as a future threat (see also Jalbert et al. 2017 for telling cases in North America and other non-African settings). It seems also that the Malthusian argument is making a come-back—too many people on too small and shallow resource bases, leading to unsustainable depletion—because the technological advances or the needed large-scale application thereof is not keeping pace with problems of global warming, climate change, high population growth, and decline of local knowledge management and capacity. So, cliché or not, the challenges to the African—and the global—environment are still there and are very real. Sheer numbers are having their impact. As Milman reports on recent research (2015): 'Humans are "eating away at our own life support systems" at a rate unseen in the past 10,000 years by degrading land and freshwater systems, emitting greenhouse gases and releasing vast amounts of agricultural chemicals into the environment.' While the evidence is not supporting an overall decline in vegetation everywhere in Africa, its nature and quality, as well as the biodiversity and ecosystems services, do suffer (cf. Brandt et al. 2017).[5] Some countries are also more hit by ecological decline and population pressure than others.[6] In almost all cases, however, societal, political, and economic factors mediate between environment and people; i.e., it is rarely appropriate to simply 'blame' local small-scale producers, indigenous/ethnic minorities, etc. for processes for the decline of their environments. Resulting processes of environmental degradation also have major, but underestimated, security implications (cf. Meier et al. 2007; Burgess 2008; AIPC 2017).

This book is a multi-disciplinary, radically 'problem-oriented' collection, considering some of the direct challenges of environmental phenomena and problems in specific African contexts. It takes a strong interest in the interactions between people, environment/ecology, politics, and (cultural and social) values in light of the debates on the 'global crisis.' There is perhaps no autonomous 'environmental' dynamics but only a human-made one: a socioeconomically, politically, and culturally embedded dynamics, leading to the ecological threats mentioned. The book rehearses a number of familiar themes in the study of the environment as known from political ecology, environmental studies and anthropology, one of them being the problem of small-scale land users being connected to global networks and markets in contexts of unequal power and state exploitation whereby local knowledge that had served societies for ages (cf. Hendry 2014) gets subverted and adaptations get destroyed. Often, as Walker noted (2005: 74), it then followed that a 'situational rationality' was created that might force people to degrade their environments and lose grip on their productive conditions. In this respect, the situation—also in Africa—has not much improved. As most case studies in this book show, such processes continue and increase the ecological–environmental problems.

A major challenge is 'how to talk about environmental problems so people, notably policy-makers and power-holders, will listen,' and this includes scholars (cf. Mann 2015), because cognizance of 'the facts' does not gear people or even governments into action by itself. How do we contribute to a new perspective on the enduring problems of African environments that both recognizes its global connections and negative interactions with wider exogenous processes—imposed changes, economic exploitation—and the potential in African societies and cultural traditions for recovery or sustainability? What is the nature of environmental change in Africa in the present era, and what is 'special to the continent'? How do local communities perceive and face the threats and visible changes in their environmental conditions? 'Doom and gloom won't save the world' (Knowlton 2017), but confronting the challenges is needed.

Myths of 'Community'?

Making the continuing problems visible and show connections is one thing. It is not only in Africa that the environmental crisis is brewing; it is intertwined with the global trends, as the studies of the IPPC and other major research institutes time and again emphasize. An important warning came out in November 2017, authored by a large group of concerned scientists (Ripple et al. 2017). But especially after the Paris COP21 conference in 2016 governments in Africa have also to take more responsibility in recognizing and managing the creeping crisis and, we assert, reconstitute the political playing fields and the 'stakeholder' lineups. It is not only a question of developing so-called effective government policies, and green energy plans to replace the fossil fuel-burning economies and massive clearance of lands, but also to involve local communities in co-managing their environment and agroecologies and proceed on *what works locally*, in functioning communities. Indeed, agroecological systems that have evolved over many ages and having achieved some balance between environment and people as well as having retained important levels of biodiversity are to be rediscovered and invested in much more. There are 'success stories' showing that local people find solutions if challenged and adapt to problems when allowed agency and when state authorities and international donors back off (cf. the remarkable studies in Zaal and Burger 2009, and the classic discussion in Leach and Mearns 1996). In particular, massive commercial agriculture to replace smallholders and agropastoralist economies is not always the solution (cf. De Schutter and Frison 2017). On the relation between agriculture and biodiversity, a study by Mendenhall et al. (2014) noted that '[t]he hospitality of the world's agricultural lands is threatened by an increasing use of chemical inputs and practices that sterilize, structurally level, and standardize plots—homogenizing and decimating biodiversity.'

Predictably, climate change will remain one of the background issues of environmental change and decline. But this book is not about that; climate change is a well-known and well-charted phenomenon the rough outlines of which are very familiar, also for Africa (cf. Klein 2015). Here, we are rather presenting a series of case studies of what is actually happening in local settings in Africa through the eyes of those concerned: People seeing that their environment changes and cuts away their livelihoods, and who often feel powerless to act against decline.

Our central idea is that the global and African growth narrative—fueled by World Bank and donor discourse as well global multinational companies and 'new partners' such as China and other resource-seeking countries—does not sit well with developments 'on the ground,' in contexts of informal urban economies or rural economies that are small-scale and still on a low technological level, but that provide livelihoods and in Africa still produce most of the food and many other products. In the context of local societies, the emerging environmental crises are most visible, starting with seemingly insignificant events and problems. Two telling examples: a case in Sudan, described by Abdalla (1985), shows that the cumulative cutting down of forest led to the destruction of a fragile ecosystem and livelihoods, due to declining rainfall and growing population. Nobody chose for it, but the community sank back into poverty and became dependent on remittances. In Cameroon, a culturally codified 'taboo system' on resource use on Mt. Oku was invalidated by government intervention in the exploitation of forest products and by harsh economic 'market logic.' The people's local knowledge system was disrespected and the cultural inhibitions on use of the 'sacred' mountain, which functioned as inbuilt conservation strategy, went into decline and the result was destruction of the ecosystem (described by Fisy 1992).[7]

Such contexts and such cases are now very numerous and cover the entire continent. 'Traditional' ideas on the environment are cut loose or invalidated in national and especially international narratives of growth—these have little eye for sensitivities, meanings, or context and are hardly rights based. Modernizing and state approaches lead to direct instrumentalization. How these problems add up and relate environmental management problems, disputes on land and resource access, and also indirectly to high population growth is what should concern us. A significant body of literature is available on them, and they need of course more in-depth historical, socio-ecological, and anthropological study—especially into agroecological systems and local environmental knowledge traditions (On which there is in fact a long tradition of research, cf. Bassett and Crummey 2003; Johnson and Hunn 2010). But policy circles seem non-responsive. The greatest disappointment in the entire environmental policy venture since the last few decades has been the World Bank, which has consistently taken the side of governments and state policy makers, and has—despite the good sounding rhetoric—never really supported the struggling local and indigenous communities and often contributed to the demise of (access to) their environmental resources (e.g., Bovard 1987; Oakland Institute 2014).

This book addresses such issues of local dilemma and response, based on situated case studies and fueled by the sense of urgency and concern on the observed decline of ecological variety, richness, and resilience. The theoretical approaches are varied but stress the underlying role of social relations, inequalities, and cultural structures that inform productive economies, policies, and the relations between state authorities (elites), private investors, and local communities of farmers, agropastoralists, and other local resource users. We are particularly interested in the idea of community-based responses and restoration efforts (cf. Leigh 2005) as they evolve in their relationship with wider processes of a political–ecological nature. Indeed, the basic underlying theoretical orientation here could be termed 'political ecological.' But the approaches taken here are interdisciplinary, with a strong representation of rural sociology, environmental studies, political economy, and social anthropology, a discipline that situates arguments and case studies in the sociocultural contexts just mentioned (cf. Haenn et al. 2016; Townsend 2009; Kopnina and Shoreman-Ouimet 2011 for path-breaking collections). Next to the fact of objectively declining natural conditions and quantity of 'resources' available on Planet Earth, it needs to be recognized that power differences, social hierarchies, and social and cultural value orientations regarding 'the economy' and politics are crucial in determining the emerging contours of the environmental crunch. That also goes for the 'communities': since at least Leach et al. (1999), we know that 'communities'—ethnic or otherwise—are often divided in their selective engagement or cooptation with regard to non-community agents. This can lead to social fragmentation and insertion of emerging 'elites' into wider economic–exploitative networks that have little business with local agroecologies and 'common resource management.'

Political Ecology

'Political' conditions are thus of inevitable importance in considering the evolution of African environments: historical trends and conditions, resilience and loss, economic predation and inequality as well as cultural codes, and deeper structures interact to reshape the environment across the continent. Well-known and plausible are the arguments by Blaikie (1985) and Watts (1983)—pioneers of the political ecology approach in the 1980s—that soil degradation and resource pressure were for a large part the result of colonial land policies. But thirty years later, one cannot keep on blaming colonialism for all ills, as current skewed and predatory patterns of resource abuse and international interference and competitive investment, buttressed by African state elites—contribute to the decline processes, aided by global capital.

There is a race between 'development' conceived in its grand scheme, top-down planning mode, and its ecological–environmental damage effects on the one hand (as numerous studies continue to reveal) and the desires/policies to counter them and do 'inclusive,' 'sustainable development for all'—as the new global, disembodied discourse on the level of UN, EU, etc., conferences has it. It is doubtful, even on the basis of environmental algorithms of resource use and predation, if a balance can be reached at all.[8] This is a global policy and governance dilemma that has as yet no solution. Africa is, unfortunately, in bigger danger here, because it chose the way of massive grand-project development, including the 'developmental corridors' and the leasing out of its lands to foreigners, mostly for short-term cash revenues for the national treasury or for private corporate interests. The local uptake of such investments is, however, still negligible, as they often remain extraneous hired labor enclaves. Africa thereby has often weak governance structures that hardly monitor the processes, do not provide sufficient legal guarantees, and do not give citizens a stake in deciding and finding local solutions or alternatives. Africa is also still in an undue dependency relationship toward the global financial institutions (World Bank, IFO, IMF) and tends to repeat a process of development—with all its problems—that took more than 200 years in modern-industrial nations. Although the latter now try to remedy the environmental damage done in the process, African governments, in this mimetic drive for 'rapid development' in the mirror of developed, wealthy countries, are paying less attention to environmental aspects and marshal global capitalist forces in their own exploitation, and even adopt them for their own national development. There is serious doubt if this is going to work, if mechanisms for sustainability and the management of ecology and population are not *built in* simultaneously. In addition, Africa's population growth is totally unique and shows little signs of abating. It is hard to see what the advantage is of population growth rates that are spiraling out of control. This complex issue—connected to religious and cultural issues and educational failure—is perhaps the most enduring challenge to things like social equity, balanced economic growth for all, political stability, and environmental sustainability.

In all these developmental efforts, there is an unquestioned focus on human activities in nature, i.e., systematically privileging humans as having the sole right and power to reshape natural conditions. Other species—plants, trees, animals, and pathogens—are the décor for utilization and exploitation. If such an observation would be made on conferences and policy meetings, the response would be: Of course! Humans are first; it's about development, eradication of poverty increasing well-being, etc. But this—in itself understandable—perspective neglects the systemic aspects of species-environment interactions and dependencies and that has led to current ecological crises.[9] Global pollution scenarios are unfortunately rather negative: the unstoppable population growth[10]; plastic waste in the oceans,[11] rising sea levels due to unprecedented ice cap melting, erosion, water scarcity, species loss: Is there any positive development? Yes, investments in solar energy, windmill energy, energy-saving devices, the rise of global consciousness on the elite level. Some countries announced plans for forest regeneration, such as Ethiopia in January 2018.[12] But globally speaking, there is little offsetting of positive and negative trends (Ripple et al. 2017; Milman 2015). Consumerist modernity has gripped humanity, and the crunch scenario is inexorably gaining ground. It is difficult not to fall back into 'apocalyptic' language about Earth's problems; it seems as if humanity is being presented the bill for unabashedly predating upon nature and not having factored it in as production cost for its activities (cf. Barbier 2014).

The studies here aim to illustrate the dilemmas and challenges of facing and managing emerging environmental crises in Africa. They often seem 'localized,' but in reality are not. The development of remedies and turnarounds is an issue of government policies but also of enabling local people, whose ecological traditions and time-tested practices must be marshaled against the new ambitious large-scale, often imposed growth schemes and commercial ventures that are a part of the problem.

We speak in the title of the 'crunch,' but we would have to add a question mark. Alarmist narratives on global environmental conditions are quite widespread, and can take over policy discourses or public perceptions, leading to indifference. Environmental alarmism is a genre of writing (cf. Middleton 2012), but we go beyond that in posing the problems anew. In Africa—among other global regions—the basic, systemic tension between the ebullient growth aims and figures for the past decade and the realities and tensions generated by this rapid (state-)capitalist-led surge have to be dealt with. The 'crunch' occurs against rapidly rising population numbers, weighing on the urban and rural landscapes and a threat that undermines several aims of the UN-enshrined 'sustainable development goals.' The pressure on the global environment is due also to the faulty, often predatory economic valuation of 'resources': As noted, the latter are still largely seen as 'externalities' that need not be costed. The paradox seems unresolvable, but this problem will hit back society and nature (cf. Gies 2017).

In particular, the drive toward large mono-crop plantations—as the 'example' of agrarian modernization—is not good news and has led to undue land degradation and rupture instead of continuity with existing (food) production systems and ecologically diverse landscapes. The craze for 'development corridors' also has predictable costs that will bounce back (cf. Laurance et al. 2015). More voices, however, seem to support the need to rethink and reorganize modern large-scale commercial agriculture and refocus on building biodiversity, respecting more ecologically embedded production methods (cf. De Schutter and Frison 2017; Nabhan 2016). The idea that only *more* modern agriculture can 'save' the world and its growing demands for food and resources is mistaken, because the preconditions for nature's reproduction and variety are being undermined. A recent example (in 2017) was the scare about the banana, a mass-produced global crop that due to its being reduced to basically one variety was under threat of decline due to a fungus (Young 2017). Biodiversity destruction is therefore a loss that—even apart from the huge aesthetic impoverishment of the world's natural landscapes and richness—makes crops quite vulnerable, and with it, food production and environments (Cardinale et al. 2012). There are clear economic dangers here (see Biodiversity International 2017). The current clearing and degradation of forests alone—especially acute in the developing world (Asia, Africa, and Latin America)—was recently estimated to contribute 15% to global greenhouse gases: '…more than all the cars, trains, planes, ships and trucks on earth' (Ricketts et al. 2010).

As noted, most chapters here theoretically follow a political ecology approach, which combines an analysis of environmental conditions with those of human social/political organization, land access rules, and power relations, that again condition, or at least impact, the access to use of environmental/ecological elements (cf. Martinez-Alier 2002: 70–71). While the concept of 'political ecology' was first used in 1935, the anthropologist Eric Wolf referred to it in a 1972 article. Since then it was taken up by geographers and many others and rapidly expanding its influence (cf. Martinez-Alier 2002: 71–74). Our use of the term (certainly in this Introduction) is inspired by the anthropological tradition, which sees the interaction of 'culture,' politics and environment (ecology) as key. The subfield of 'cultural ecology' (since J. Steward in the 1950s) always stressed the cultural context and its impact on the definition and use of 'the environment' in relation to power structures, but focused on a local, working and often 'quantifiable' socio-ecological system. This 'traditional' school of cultural ecology yielded a new perspective and produced magnificent studies like Roy Rappaport's *Pigs for the Ancestors* (1984). Today, this traditional approach is much more over-layered by politics and political geography, so much so that the environment and its actual biophysical characteristics and processes often are relegated to secondary place. This urged Vayda and Walter (1999) to write a critique in their paper 'Against political ecology,' although their case may be overstated: Walker (2005) suggested that in much work the biophysical aspects are *not* neglected but integrated into political analyses. Still, the argument that all problems are of a primarily political nature and to be remedied by corrective action while the environment offers possibilities and resources for all is misleading. A refocus is perhaps needed on the limits of the global physical environment and the facts of growing masses of people and rising consumer standards. Here, we would plead for a re-emphasis on the *cultural* ecology of production and environmental conditions: How are human–environment interactions embedded in and dependent on cultural narratives and rules, and how do these contribute to survival and sustainability? How did agroecologies, agropastoral systems, etc. survive? The call for a 'new cultural ecology' has been made before (cf. Kottak 1999; Haenn et al. 2016: 8), in the sense of 'finding practical solutions to environmental problems' (ibid.). We mean here a theoretical and empirical re-orientation on the bio-cultural traditions of production and environmental interaction that local societies and producers in Africa demonstrate and that reveals intricate connections between society, culture, and environment that go against a direct instrumentalization and commercialization turning 'nature' into a commodity. As already mentioned, various studies in the book show that the massive development ventures and land commercialization processes on the globe start with serious neglect of context and culture and impose their 'rational' economic management and commodification routines on environments that ill-tolerate them in the long term. Vital ecosystems elements are thus removed, and with them, the eco-knowledge complexes that evolved over time. Actual environmental decline (and recovery) processes are better reconnected with social, cultural, and political factors, and not by seeing them as only derivative of the latter.

Needless to say, this book has deeply shared concerns about environmental issues in Africa but no unity of perspective and approach, not even geographical. Reviewers will no doubt note this. But we make no apology for it: The collection reflects the surprising diversity of problems, the multifarious challenges to society and environment in Africa, and also the different registers in which the problems are discussed, from literature to state policy-making to post-cyclone emergency aid. The case studies go on to show that from many perspectives—e.g., economic (Nasongo et al., Buckner, Djohy et al., Jensen), political–ecological (Abbink, Gabbert, Tsuruta), sociocultural (Eze, Llopis, Tesi)—the environmental threats, if not the coming *crunch*, are quite real and are produced in problematic socioeconomic practices, often in the context of deep political problems caused by doubtful elite power politics and a global resources grab fueled by foreign investor from the richer countries in Asia and in Europe or the USA. The chapters all point to a similar underlying message: The environmental resources in Africa are in relative and absolute decline in many places, people lose grip on their environment, and 'sustainable resource use' is elusive, despite the expressed rhetorical commitment of governments to the SDGs in fancy global meetings. And it is not productive to deny the fact that historically unprecedented population growth goes on to drive this, unchecked by sociocultural norms of restraint or serious policy measures. Here, we touch again on the neo-Malthusian problematic. While the classical nineteenth-century Malthusian approach, on the 'exponential' growth of population vs. the arithmetical growth of resources and food production, was refuted by the unexpectedly rapid technological–agronomical advances of the last century and a half, the population growth issue has returned in full force, due in part to (a) the post-World War II development aid enterprise that put priority on improving health care, without thinking of the consequences of hugely increased numbers, (b) failing industrialization and modernization of the poor countries in the developing world, and (c) the high level of conflict, violence, and societal destruction in the same countries, partly a consequence of colonial or dictatorial rule that destroyed much of their sociopolitical fabric. It is, however, remarkable, that of all developing countries only China chose to apply an official state population control policy. This policy no doubt has strongly contributed to its fabulous economic growth since the late 1980s.

Hence, the neo-Malthusian challenge is back. Not in all countries in the developing world the 'Boserupian' argument (Boserup 1965) applies—that rapid population growth can lead to intensification and innovation in the agrarian sector; this only holds under certain conditions. Africa in particular is not showing a dramatic decline in demographic growth (no 'transition'), defying a general trend that when countries develop (educationally, economically, and socially) the rates tend to fall. In fact, Niger has the largest population growth rate ever recorded, with 3–4% annually.[13] One can speak of a 'negative feedback loop' process that bodes rather ill for the African continent. It is corroborated in most recent research (Ward 1994; Ceballos 2015; Ripple et al. 2017, etc.). As Moses Tesi also hints at this in his chapter below, discussing West Cameroon: 'No amount of improvement in agriculture is going to absorb the region's large population on the area of land available' (Chapter 7). In this book, we do not follow the old vision of over-population and population growth as *the* cause of disastrous environmental effects. But to deny any relationship between the two is simply incorrect. Sheer numbers speak for themselves. As always, these numbers are 'mediated' by cultural, religious, legal, and political prisms, and various elites seem to want to tell people what to do, not to get enlightened about rights to health and female empowerment, and exercise family and mind control. But the effect is continued high and unaddressed population growth nevertheless. Many countries in Africa fall under this: e.g., next to Niger, also Ethiopia, Sudan, Somalia, Nigeria, Burkina Faso, or Uganda.[14] As land and resources do not grow but can only be used more intensively and exhaustively, population growth will keep spiraling, without quantum leaps in agro-technological investments and urban-based industrialization able to absorb people. It goes without saying, this crisis is already there, and visible in the mass migrations and population displacements. The causal paths and the motivational context of human reproductive behavior need to be studied carefully in each country and will show the combination of factors responsible. Clear is that constraining women's rights, semi-forced early marriages (child marriages), lack of freedom to move, and lack of education and employment all have a negative impact. Women burdened early with many children are socially neutralized and kept subservient, in turn holding their kids and families in relative poverty.

The Case Studies

The first three essays in the book are the most rigorously empirical, based on excellent fieldwork and describing developments on the ground with a wealth of closely researched data. The same goes for Tadasu Tsuruta's chapter later in the book. These are telling accounts of how difficult the situation becomes for local producers, through no free choice of their own and confronted by new schemes and processes of investment, commercialization, and development that have both foreseen and unforeseen effects.

Margaret Buckner describes how, starting in the 1960s and 1970s, but especially since the 1990s, cashew production has boomed in Guinea Bissau, encouraged by the World Bank, the Guinea Bissau government, and European and American development agencies as well as some NGOs, and experts everywhere. This support was in keeping with development discourse and the neoliberal growth narrative. But as cashew production increased, rice production declined: rather than growing their own rice, people grew cashews to exchange for rice *imported* from Southeast Asia. Smaller rice harvests and dependence on world cashew prices thus led to food insecurity and social instability. Furthermore, cashews took over local ecosystems, and once common plants and animals disappeared. In other words, a self-sufficient, locally controlled, sustainable, environmentally friendly subsistence economy was turned into an unsustainable, monoculture economy depending on fluctuating global markets. The shift brought about changes in social organization, gender relations, and religious activities and is wreaking havoc on the environment. Who is helped by this?

The study by Serena Nasongo et al. looks at the changing status of the Ombeyi wetlands in Kisumu County, Western Kenya, an area used for crop production and grazing livestock, as well as providing ecological services and supporting local livelihoods through resources such as papyrus products, fish, and water supply. The increased demand for agricultural commodities put pressure on the Ombeyi wetlands due to its high agricultural productivity at relatively low costs. The authors describe the pressure building up due to commercialization of agriculture (rice, sugar cane), the introduction of new and profitable but environmentally disadvantageous crops (such as arrowroot) and the resulting problems. Due to tedious work and the high cost of operation in rice production and poor markets for rice, some farmers have shifted to growing the other crops with better markets (sugarcane, arrowroots). But arrowroot cultivation led to water use conflicts, declining rice yields and low soil fertility due to poor agronomic practices. Based on team-fieldwork findings, the chapter demonstrates the need for *integrating* sustainable wetland management and agriculture in Ombeyi regarding wetland hydrology, ecology and socioeconomics in order to avert the degradation of the wetlands and maintain the delicate balance between wetland ecological processes and local livelihoods in the Ombeyi watershed.

In the next chapter, Djohy et al. use the perspectives of political ecology and of science and technology studies to highlight how sociotechnological change over recent decades has influenced cultivation and livestock keeping in northern Benin. The agricultural technologies introduced are negotiated and adapted in use to shift power relations to the benefit of some rural actors over others. The authors trace labor-saving technologies such as tractors and especially herbicides and analyze their effects in light of ongoing uses made by both smallholder farmers and pastoralists. Results show a 'renegotiation' of herbicides from weeding technologies to weapons for contestation over land and natural resources. Herbicides allow farmers to expand their crop fields in a labor-constrained context and to lay claim on or maintain ownership of land. However, the use of chemicals has reduced the extent of rangelands by polluting grazing lands and water resources and has increased farmer–herder conflicts. Many herders left the region for neighboring countries such as Togo and Ghana, with economic consequences because of decreased livestock and milk marketing.

In Stig Jensen's chapter, a quite different subject is treated: eco-tourism. He discusses both livelihood opportunities for rural societies located near natural reserves and the possible conservation of biodiversity in an internationally important protected area in light of growing Tanzanian *urban-based* demands for and investments related to ecotourism, using Mikumi as case. This park in recent years saw an increase in both Dar es Salaam-based ecotourists coming as well as growing external investments in small-scale tourism facilities. It affected both communities in the area and the protected area (mainly key species) in different ways. The issue is compelling and possibly challenges conventional wisdom based on the assumptions that the combination of nature conservation and ecotourism both enhance poverty in rural societies in Africa and improve environmental conditions especially for flagship biodiversity in protected areas.

In a more interpretive essay, Abbink revisits a study of the Suri people in Southern Ethiopia that he followed for over two decades, and notes a demise of Suri livelihoods and culture in the wake of development offensives of the Ethiopian state. It is a survey of how a local agropastoral society in Southwest Ethiopia conceptualizes and experiences changing 'resource use' and of spiraling environmental problems in a time of state-imposed agrarian development schemes. The process of state-induced political-economic change since the 2000s, including monocrop plantation agriculture and resettlement, shows no investment in local economies and diminishes local environmental stability and biodiversity. Local ethnic groups—both agropastoralists and smallholder peasants—are under pressure due to shrinking resource availability, influx of tens of thousands on newcomers and stronger state presence, including in military form. Local peoples seek new opportunities to connect, both to local and national arenas of power, but are so far no successful in this. The 'political ecology' approach followed here reveals state economic and political expansion leading to environmental impoverishment, subversion of local livelihoods, and a de-emphasizing of ethnic and civic group rights—a situation that could only be remedied by state reinvestment in local economies.

Moses K. Tesi's chapter on Cameroon's Western Region discusses the evolving environmental tensions building up in this smallholder agricultural region, notably in the Bamileke districts. The region's high population density and small geographic space that forms the basis of intensive agrarian-economic activities for more than 90% of the population are turning it into an emerging environmental bombshell waiting to explode. The national government has stimulated urbanization due to its creation of new administrative divisions and 'services,' etc., but developed no policy to deal with the induced population growth. Only stimulation of non-agricultural investments can relieve the pressure on the land and the ecosystem in place.

Tadasu Tsuruta's chapter explores, on the basis of meticulous fieldwork, the transformation of agropastoralism in semi-arid Tanzania. As the semi-nomadic way of 'traditional' agropastoralism has become no longer feasible today, the bulk of the rural population resides permanently in densely populated villages, and many villagers ceased to keep livestock, having come to depend on agriculture, which is highly unstable in a semi-arid climate. At the same time, villagers have increasingly engaged in other income-generating measures, many of which are dependent upon dwindling forest resources. Land is now concentrated in the hands of a small number of wealthy cattle owners, due to a growing scarcity of grazing land. A shortage of rangeland is further exacerbated by the rapid expansion of commercial maize farms, partly enhanced by the agricultural modernization policy of the Tanzanian government, which allows easier access to imported tractors and agricultural loans. Expansion of farmlands has put a considerable amount of pressure on the environment, leading to rapid deforestation and active soil erosion.

In his study of Behompy village, Madagascar, Jorge Llopis shows the complexity of the challenges that environmental governance and poverty alleviation objectives face in the context of changing climatic conditions, based on a case study from a protected area in southwest Madagascar. Temporary protection status was granted in 2008, with the objective of conserving the spiny forest in the region while allowing its sustainable exploitation by rural inhabitants. The area has been suffering one of the highest rates of deforestation in the country while receiving the recurrent impact of droughts and cyclones that further reduces the alternatives that rural populations have to pursue sustainable livelihood strategies. The research deepens into the diverging narratives deployed by the different actors involved in the management and use of the natural resources to enhance our understanding of the social–ecological processes taking place in an area historically bypassed by development projects. With unabated high population growth rates and acute impoverishment of rural communities in many African countries, neo-Malthusian explanations for the perceived environmental degradation trend—focusing only on 'population growth'—are gaining momentum in the international developmental and environmental policy-making spheres but Llopis is not convinced of their plausibility.

A unique approach is provided by Norbert O. Eze in his study of the literary imagination of climate change and environmental problems in Nigeria, much present in the public consciousness. Dramatist Greg Mbajiorgu's play has posed this major theme of the ecological 'crunch,' evident in phenomena of drought, desertification, oil spillage, and now unprecedented flooding, in his influential play *Wake Up Everyone*. These problems are as important for ordinary Nigerians as 'poor governance' and the 'Boko Haram' insurgency. The diverse ecological challenges, the plays suggests, necessitate a review, change or adaptation of ways of living, dramatizing the immediate problems posed by the environment. This eco-drama by Greg Mbajiorgu and the way how he articulates and problematizes the challenges militating against ecologically sound policies in the country is analyzed in-depth in this chapter.

Finally, Echi Gabbert closes with a strongly reflexive and 'experimental' chapter, giving an interpretive turn to the discourse and dilemmas of environmental threats and 'development' in Africa. In response to forms of new land use in the agropastoral regions of Ethiopia, many actors have different views on how to handle the same territories. In light of the rapid pace of social and environmental change induced by fast-track development plans in times of climate change, it is urgent to take a step back and reflect on the history, ecology, and future of the people and the land that are in transition. Inspired by the late Zygmunt Baumann, she first pictures a global 'dystopian' view, then a 'utopian' view of the Ethiopian context, and presents an ethnographic example from Southern Ethiopia to outline the need for alternative solutions for present challenges of climate and food insecurity and finally to redefine the notion of 'experts' on land.

Conclusions: A 'New Cultural Ecology'?

Conclusions are not definitive here but could fall into two categories: scientific and political. Ecological–environmental conditions in Africa are worrying and are impacted strongly by the contradictions between the grand development narratives of states, donors, and international organizations and the more local, agroecological knowledge-based or bio-cultural heritage systems, which are subverted without having been examined on their merits or invested in. Agropastoralism is one of the examples; integrated agroecological cultivation (cf. Getachew 2014) or river-bank (flood retreat) cultivation are others. Not all (agro-)pastoralist production systems and societies have a future; some are disorganized and pose a security threat (see the alarming report by AIPC 2017), but most of them need investment, not neglect.

Our contention is that after the predominance of political ecology in environmental studies in Africa, we might need to refocus again more on *cultural ecology*; i.e., a new cultural ecology informed by the tradition of anthropology that recognizes the interaction of varying conceptions of environment and production of food and resources, and that accords more space and value to local, culturally informed patterns/traditions of interaction with the natural environment, which reflect adaptive solutions.

As to the politics of development, the following could be remarked. Fresh 'political' analyses of the mounting problems and their connections—between politics, culture (including religious thought), and economy—are needed, and a political turn-around across the globe is desired. It will probably not happen except in specific countries and locations.[15] Even if the scientific evidence on ecological decline and the rates of humans suffering are plain to see and copiously documented in hundreds of studies, governments and regimes are of course not everywhere beholden to 'improving the environment, sustainability,' etc., but rather to the GDP 'development' of their country. The one does *not* equate the other: development, notably in Africa, is primarily a state and power-building strategy: to generate more income for the state elites and the regimes in place, intentionally or not at the expense of society and nature. As we know from a growing body of literature in the wake of the Th. Piketty debate, socioeconomic *inequality* worldwide is also increasing, along with repressive state surveillance capacity—and thwarts authentic social protests, alternative narratives, and existing local or indigenous economic systems, thus impeding a new sustainable economy based on the circularity of resource use. If democratic and accountable structures of decision-making and deliberation are not further developed, the environmental problems will likely not be solved. In Africa, the democratic deficit is very serious and still a great hindrance to good environmental governance (cf. already Bond et al. 2002). And if, in addition, the current (2017) President of the most technologically and economically advanced power on the globe, the USA, denies the findings of scientific research on climate change, ecological decline, and social inequality, then we have a further setback. Rational policy choices and international agreements are thus made hostage to uninformed elite rule, ignorant consumerist ethics, and aimless hedonism.[16] The effects were already seen in the negligible US role in the COP23 conference climate problems in Bonn, Germany, in November 2017. Next to the USA, other countries, like China and India, do their best to contribute to the environmental crunch. There is ongoing massive fossil fuel burning in the name of 'development', excessive and unregulated population growth, unrelenting destruction of natural settings and biodiversity[17] for misguided large-scale land-clearing, and dam-building ventures, and last but not least religious discourse if not bigotry intimidating government policies, closing minds, and forbidding rational debate on shared problems.[18] In the domain of fossil fuel burning China stands out, with its still massive use of coal, which caused a 2% *increase* in global carbon emissions in 2016 alone. We simply have the global phenomenon of growing consumption of energy and fossil fuels as countries develop, leading to an inexorable emission growth in the coming decades, as more wealthy elites and upper (middle) classes emerge worldwide (cf. Eisenstein 2017). But even in highly industrialized countries like Germany, with a government nominally committed to reversing climate change, there is a reluctance to reform the coal and brown-coal fueled industries.

In short: the phenomenon of 'development' is often working against sustainability (in Africa and elsewhere), which carries dangers to aggravate the ecological crises. The challenge described by Merle Sowman and Rachel Wynberg (2014: 333f.) still stands: improving governance for justice and sustainability is an African and global necessity. A new global policy discourse and developmental regime are needed, *not* to replace well-working and productive systems of modern agriculture in the developed world, but developing new economic spaces with a concern for integral environmental policies. These can be informed by the *new cultural ecology* that recognizes the political dimensions but is not overdetermined by them, looks at actual practices of problem-solving interaction of humans and environment (cf. Marshall 1995), and takes bio-cultural heritage and agroecological complexes seriously. Bridging the gap between global growth narratives and the unfavorable local environmental realities that we see across Africa and the globe is the way to more sustainability and equity. An appeal to the underlying *economic* problems of the ecological crunch ahead of us can help to 'make people listen' (Mann 2015). The findings of environmental science and from cultural–ecological studies—as partly reflected in this book—need to be taken more seriously and converted into policy.

Notes

1. See the 2014 IPPC summary synthesis report (http://www.ipcc.ch/pdf/assessment-report/ar5/syr/AR5_SYR_FINAL_SPM.pdf), and the complete report (*Climate Change 2014 Synthesis Report*), (http://www.ipcc.ch/pdf/assessment-report/ar5/syr/SYR_AR5_FINAL_full_wcover.pdf, accessed December 5, 2017). Or just read this, on the Larsen-C iceshelf that recently detached: K. Grovier, The crack that's redrawing the world's map, BBC, May 5, 2017 (www.bbc.com/culture/story/20170505-the-crack-thats-redrawing-the-worlds-map, accessed May 7, 2017). See also Vidal 2009, on the warning on Africa by then chief scientist of the UK, G. Conway.
2. E.g., visualized in this film: https://www.youtube.com/watch?v=nKieqi9BNkc, with data on the Lake Victoria area (1960–2015), and the Tai National Park in Côte d'Ivoire (UNEP data).
3. See http://www.worldometers.info/world-population/africa-population/ (accessed January 15, 2018). Also Engelman (2016a, b).
4. See the *Global Environmental Outlook (GEO-6): Regional Assessments*, cited in 'Rate of environmental damage increasing across the planet

but there is still time to reverse worst impacts if governments act now, UNEP assessment says,' May 19, 2016 (www.unep.org/newscentre/Default.aspx?DocumentID=27074&ArticleID=36180&l=en accessed August 10, 2016).
5. They noted '….two different trends in land area with woody cover for 1992–2011: 36% of the land area (6,870,000 km^2) had an *increase* in woody cover largely in drylands, and 11% had a decrease (2,150,000 km^2), mostly in humid zones. Increases in woody cover were associated with low population growth, and were driven by increases in CO_2 in the humid zones and by increases in precipitation in drylands, whereas decreases in woody cover were associated with high population growth.' (From the abstract).
6. For one study on Nigeria, see Onwuka (2006). See also Nigeria Health Watch (2016).
7. Also, compare the collection edited by Sheridan and Nyamweru 2008.
8. As I wrote this in August 2017, the Brazilian government of President F. Temer decided to open up a large national forest reserve in the north of the country, with irreplaceable ecosystems and biodiversity value—for exploitation by the country's predatory agro-capitalist class for gold mining, mono-crop plantations, and timber logging. The future is guaranteed: destruction and pollution of ca. 45,000 km^2 of Amazon forest (for which Brazil has a global responsibility) as well as displacement and marginalization of local Indian populations will follow. For short-term political gain, Temer wanted to buttress his embattled domestic political position and needed the support of the business class—and was prepared to sacrifice his country's environmental riches. Similar examples are found in Africa.
9. See also the 'ecological footprint' discussion, https://www.footprintnetwork.org/our-work/ecological-footprint/#worldfootprint.
10. See http://www.worldometers.info/world-population/. Somewhat frightening.
11. See BBC, 'Seven charts that explain the plastic pollution problem,' at http://www.bbc.com/news/science-environment-42264788.
12. See https://africareportonbusiness.com/2018/01/17/ethiopia-ambitious-plan-to-restore-forest-landscape-revealed-2/ (accessed January 29, 2018). Ethiopia currently loses ca. 92,000 ha. per year.
13. In 2017, it was 3.99 per woman of childbearing age. http://countrymeters.info/en/Niger, accessed January 08, 2018.
14. In this context, it is regressive and self-defeating that the USA has recently (in 2017) decided to stop its support for female sexual health and education and family guidance programs in Africa, out of some typical American puritanical–religious agenda.

15. If only because of corruption getting in the way in many developing countries; cf. Sarah Chayes in her major book on the destructive role of corruption worldwide (2009: 186, 207). See also: 'Bonn chance. UN climate talks finesse details as disconnect grows between rhetoric and real-world trends,' *Nature* 551: 413 (2017), and 'Corruption a scourge Africa can do without: UNECA deputy executive,' *Addis Standard*, January 22, 2018 (http://addisstandard.com/news-corruption-a-scourge-africa-can-do-without-uneca-deputy-executive/, accessed January 23, 2018).
16. Needless to say, there is resilience in US society and its vast scientific infrastructure, but it was dealt a major blow by the Trump presidency, and without new national science-based policies no progress can be made.
17. Since earlier 'warnings' of biodiversity decline (such as Ward 1994; Chivian and Bernstein 2008; Pimm et al. 2014), not much has changed in international policies beyond rhetorical statements as to its importance.
18. For example, a 2010 survey conducted in 10 African countries revealed this majority view: 'God, not global emissions, is to blame for climate change.' (See: http://www.greenbeltmovement.org/node/139, accessed May 12, 2014). But this attitude is widespread in the USA as well (https://climatechangedispatch.com/pew-survey-most-americans-think-global-warming-isn-t-a-serious-problem/, accessed December 20, 2017).

References

Abbink, J. (2017). Stemming the tide? The promise of environmental rehabilitation scenarios in Ethiopia. In W. van Beek, J. Damen, & D. Foeken (Eds.), *The face of Africa. Essays in honour of Ton Dietz* (pp. 185–198). Leiden: ASCL.

Abdalla, I. (1985). The *killer axe*: Farming and deforestation in a fragile ecological system in Kordofan. *Northeast African Studies, 7*(1), 59–65.

African Investigative Publishing Collective (AIPC). (2017). The war for grazing lands in Africa. https://www.zammagazine.com/chronicle/chronicle-31/494-greener-pastures. Accessed November 9, 2017.

Barbier, E. B. (2014). Account for depreciation of natural capital. *Nature, 515,* 32–33.

Bassett, T. J., & Crummey, D. C. (Eds.). (2003). *African savannas: Global narratives and local knowledge of environmental change*. Oxford: James Currey and Portsmouth, NH: Heinemann.

Biodiversity International (CGIAR). (2017). *Mainstreaming agrobiodiversity in sustainable food systems. Scientific foundations for an agrobiodiversity index*. Rome: Biodiversity International.

Blaikie, P. (1985). *The political economy of soil erosion in developing countries*. London and New York: Longman.

Bond, P., et al. (2002). *Unsustainable South Africa: Environment, development and social protest*. London: Merlin Press and Pietermaritzburg: University of Natal Press.

Boserup, E. (1965). *The conditions of agricultural growth. The economics of agrarian change under population pressure*. London: George Allen and Unwin.

Bovard, J. (1987). The world bank vs. the world poor. *Cato Policy Analysis*, no. 92. https://www.cato.org/publications/policy-analysis/world-bank-vs-worlds-poor.

Brandt, M., et al. (2017). Human population growth offsets climate-driven increase in woody vegetation in sub-Saharan Africa. *Nature Ecology & Evolution*. https://doi.org/10.1038/s41559-017-0081.

Burger, K., & Zaal, F. (Eds.). (2009). *Sustainable land management in the tropics: Explaining the miracle*. London and New York: Routledge.

Burgess, S. F. (2008). Environment and human security in the Horn of Africa. *Journal of Human Security, 4*(2), 37–61.

Cardinale, B. J., et al. (2012). Biodiversity loss and its impact on humanity. *Nature, 486*, 59–67.

Ceballos, G., et al. 2015. Accelerated modern human-induced species losses: Entering the sixth extinction. *Science Advances (AAAS)*, June 19, 2015. https://doi.org/10.1126/sciadv.1400253.

Chayes, S. (2015). *Thieves of state. Why corruption threatens global security*. New York: W.W. Norton.

Chivian, E., & Bernstein, A. (Eds.). (2008). *Sustaining life: How human health depends on biodiversity*. Oxford and New York: Oxford University Press.

De Schutter, O., & Frison, E. (2017). Modern agriculture cultivates climate change—We must nurture biodiversity. *The Guardian*, 9 January 2017.

Eisenstein, M. (2017). The needs of the many. *Nature, 551*, S142–S144.

Engelman, R. (2016a). Africa's population will soar dangerously unless women are more empowered. *Scientific American*. https://www.scientificamerican.com/article/africa-s-population-will-soar-dangerously-unless-women-are-more-empowered/. Accessed October 11, 2017.

Engelman, R. (2016b). Six billion in Africa. *Scientific American, 314*(2), 56–63.

Getachew, S. (2014). *The nexus of indigenous ecological knowledge, livelihood strategies and social institutions in midland Gedeo human-environment relations* (PhD dissertation, Social Anthropology, Addis Ababa University, Addis Ababa).

Gies, E. (2017). The real cost of energy. *Nature, 551,* S145–S147.

Fairhead, J., & Leach M. (1996). *Misreading the African landscape. Society and ecology in a forest-savanna mosaic.* Cambridge: Cambridge University Press.

Fisy, C. (1992). The death of a myth system: Land colonization on the slopes of Mount Oku, Cameroon. In R. Bakema (Ed.), *Land tenure and sustainable land use* (pp. 13–20). Amsterdam: Royal Tropical Institute.

Haenn, N., Wilk, R., & Harnish, A. (Eds.). (2016). *The environment in anthropology: A reader in ecology, culture, and sustainable living* (2nd ed.). New York and London: New York University Press.

Hendry, J. (2014). *Science and sustainability. Learning from indigenous wisdom.* New York and Basingstoke, UK: Palgrave Macmillan.

Jalbert, K., et al. (2017). Introduction: Confronting extraction, taking action. In K. Jalbert, A. Willow, D. Casagrande, & S. Paladino (Eds.), *ExtrACTION: Impacts, engagements, and alternative futures* (pp. 1–13). Abingdon and New York: Routledge.

Johnson, L. M., & Hunn, E. S. (Eds.). (2010). *Landscape ethnoecology. Concepts of biotic and physical space.* New York and Oxford: Berghahn.

Kaplan, R. D. (1994, February). The coming anarchy. *The Atlantic Monthly.* www.theatlantic.com/magazine/archive/1994/02/the-coming-anarchy/304670/. Accessed September 6, 1998.

Klein, N. (2015). *This changes everything. Capitalism vs. the climate* (1st ed., 2014). Harmondsworth: Penguin Books.

Knowlton, N. (2017). Doom and gloom won't save the world. *Nature, 544,* 271. https://www.nature.com/news/doom-and-gloom-won-t-save-the-world-1.21850.

Kopnina, H., & Shoreman-Ouimet, E. (2011). *Environmental anthropology today.* London: Routledge.

Kottak, C. P. (1999). The new cultural ecological anthropology. *American Anthropologist, 101*(1), 23–35.

Laurance, W. F., et al. (2015). Estimating the environmental costs of Africa's massive 'development corridors'. *Current Biology, 25,* 1–7.

Leach, M., & Mearns, R. (Eds.). (1996). *The lie of the land. Challenging received wisdom on the African environment.* London: The International African Institute.

Leach, M., Mearns, R., & Scoones, I. (1999). Environmental entitlements: Dynamics and institutions in community-based natural resource management. *World Development, 27*(2), 225–247.

Leigh, P. (2005). The ecological crisis, the human condition, and community-based restoration as an instrument for its cure. *Ethics in Science and Environmental Politics, 3*, 3–15 (Online).

Mann, C. C. (2015). How to talk about climate changes so people will listen. *The Atlantic Monthly*, September 2014.

Marshall, P. (Ed.). (1995). *Nature's web: Rethinking our place on earth*. New York: Paragon House.

Martinez-Alier, J. (2002). *The environmentalism of the poor. A study of ecological conflicts and valuation*. Cheltenham, UK and Northampton, MA: Edward Elgar.

Meier, P., Bond, D., & Bond, J. (2007). Environmental influences on pastoral conflict in the Horn of Africa. *Political Geography, 26*, 716–735.

Mendenhall, C. D., et al. (2014). Predicting biodiversity change and averting collapse in agricultural landscapes. *Nature, 509,* 213–217.

Middleton, G. D. (2012). Nothing lasts forever: Environmental discourses on the collapse of past societies. *Journal of Archaeological Research, 20,* 257–307.

Milman, O. (2015, January 15). Rate of environmental degradation puts life on earth at risk, say scientists. *The Guardian*. http://www.theguardian.com/environment/2015/jan/15/rate-of-environmental-degradation-puts-life-on-earth-at-risk-say-scientists. Accessed July 4, 2017.

Nabhan, G. P. (Ed.). (2016). *Ethnobiology for the future. Linking cultural and ecological diversity*. Tucson: University of Arizona Press.

Nigeria Health Watch. (2016). *The elephant in the room: Nigeria's population growth crisis*. December 21, 2016. http://nigeriahealthwatch.com/the-elephant-in-the-room-nigerias-population-growth-crisis/. Accessed January 14, 2018.

Oakland Institute. (2014). *Unfolding truth. Dismantling the world bank's myths on agriculture and development*. Oakland, CA: Oakland Institute.

Onwuka, E. C. (2006). Another look at the impact of Nigeria's growing population on the country's development. *African Population Studies, 21*(1), 1–18.

Pimm, S. L., et al. (2014). The biodiversity of species and their rates of extinction, distribution, and protection. *Science, 344*(6187). https://doi.org/10.1126/science.1246752.

Rappaport, R. A. (1984). *Pigs for the ancestors. Ritual in the ecology of a New Guinea people* (2nd ed.). New Haven and London: Yale University Press.

Ricketts, T. H., et al. (2010). Indigenous lands, protected areas, and slowing climate change. *PLoS Biology, 8*(3). https://doi.org/10.1371/journal.pbio.1000331.
Ripple, W. J., et al. (2017). World scientists' warning to humanity: A second notice. *Bioscience*, November 2017. https://doi.org/10.1093/biosci/bix1250.
Sheridan, M. J., & Nyamweru, C. (Eds.). (2008). *African sacred groves. Ecological dynamics and social change.* Oxford: James Currey.
Sowman, M., & Wynberg, R. (Eds.). (2014). *Governance for justice and environmental sustainability: Lessons across natural resource sectors in sub-Saharan Africa.* London and New York: Routledge.
Tiffen, M., Mortimore, M., & Gichuki F. 1994. *More people, less erosion: Environmental recovery in Kenya.* Nairobi: ACTS Press and London: ODI.
Townsend, P. K. (2009). *Environmental anthropology: From pigs to policies* (2nd ed.). Long Grove, IL: Waveland.
Vayda, A. P. & Walters, B. B. (1999). Against political ecology. *Human Ecology, 27*(1), 167–179.
Vidal, J. (2009, October 29). Climate change will devastate Africa, top UK scientist warns. *The Guardian.*
Walker, P. (2005). Political ecology: Where is the ecology? *Progress in human geography, 29*(1), 73–82.
Ward, P. (1994). *The end of evolution. On mass extinctions and the preservation of biodiversity.* New York: Bantam Books.
Watts, M. (1983). *Silent violence: Food, famine and peasantry in Northern Nigeria.* Berkeley: University of California Press.
Wolf, E. (1972). Ownership and political ecology. *Anthropological Quarterly, 45*(3), 201–205.
Young, S. (2017, September 18). Bananas could face extinction due to spread of deadly fungus. *The Independent.*

2

Cash for Cashews: Does It Add Up?

Margaret L. Buckner

Introduction

International entities the world over are striving to "develop" poor countries by encouraging people to grow cash crops for export, thus bringing them into the global economy. Gross domestic product (GDP) in many countries has increased, but are villagers better off? What are the true economic, social, and environmental costs of cash crops? Though economists measure the revenues that a crop like cashews can bring to a country, the effects of the crop's exponential growth in villages have rarely been examined. "Changes induced by the cashew boom at social and environmental levels are yet to be analyzed and understood" (Catarino et al. 2015). This chapter attempts to do that by examining the effects of cashew production in a Manjako village in

M. L. Buckner (✉)
Emerita, Missouri State University, Springfield, MO, USA
e-mail: mbuckner@missouristate.edu

© The Author(s) 2018
J. Abbink (ed.), *The Environmental Crunch in Africa*,
https://doi.org/10.1007/978-3-319-77131-1_2

northwest Guinea-Bissau, following in the footsteps of Lundy (2012) who studied the Nalu in the southern part of the country, and Temudo and Abrantes (2014) who focused on the Fula and Mandinga in the central and western regions.

Over the last twenty-five years,[1] I have witnessed the rise of cashew production and the corresponding loss of rice fields—quite visible on Google Earth—in Caio, northwestern Guinea-Bissau. The village has visibly changed from a relatively egalitarian population to one with haves and have-nots. Since young men have other options for earning cash and many no longer see their future as rice farmers, they no longer depend on the lineage elders for parcels of rice fields to cultivate, which leads to the erosion of lineage structure. Women, in particular, are affected, since they have fewer options to earn wages than men do. Religious practices are also being transformed, since unequal opportunity and resources lead to more accusations of spiritual wrong-doing and costlier compensations. Bush is being replaced by cashew orchards, which is having dire environmental consequences.

Cashews, native to Brazil, were introduced in Guinea-Bissau by the Portuguese colonial government. "Sarmiento Rodrigues… appointed Governor of Guinea in 1945…urged diversification of exports because of the vulnerability of rubber, palm oil and groundnuts to world competition. … Each post was to plant ten hectares of cashew and other fruit trees every year" (Galli and Jones 1987, 35; see also Temudo and Abrante 2014; Catarino et al. 2015). But it was not until the 1980–1990s that cashew production in Guinea-Bissau really took off (Fig. 2.1), with the help of international "aid" agencies. For example, "[t]he United States Government, through its Agency for International Development (USAID), is investing a further $300,000 in Global Development Alliance (GDA) funding to work with the private sector to bolster Guinea-Bissau's cashew production through training in quality control and marketing" (USAID 2005). The crop now provides the huge majority of state revenues. According to the Guinea-Bissau Minister of Economy and Finance, "the cashew campaign is the main source of income of our farmers and the main export commodity of Guinea-Bissau, and about 90 percent of our exports are cashew nuts tremendously weighting in our GDP" (UNIOGBIS 2016).

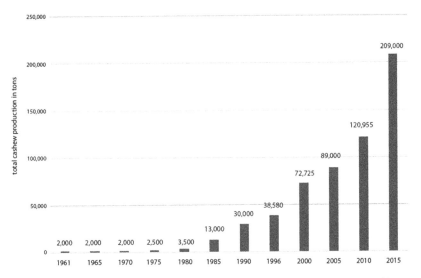

Fig. 2.1 Production of cashews (with shells) in Guinea-Bissau in tons (*Source* FAO)

Rice Farming

The approximately 9000 inhabitants of the Manjako kingdom of Caio, in the Cacheu region in northwestern Guinea-Bissau, live along a four-mile stretch that includes a central area where administrative offices and market are located and homesteads next to the rice fields that follow the coastal mangroves.

The coastal area of West Africa from the Gambia down to Sierra Leone is edged with mangroves pierced with tidal inlets that meander inland for miles; saltwater surges far up these estuaries. Rivers flow inland at high tide and then reverse flow as the tide goes out. Over the centuries, the peoples living along the mangrove belt have developed an extremely productive system of wet rice cultivation (Pélissier 1966; Linares 1992). Dikes are built around broad swaths of mangrove to hold back ocean water; they must be high enough and strong enough to withstand high tides. During the rainy season, the dikes hold in the rainwater, which floods the fields and enriches the soil with organic matter it picks up along the way. The mangrove is thus converted into very productive rice fields. However, the work is very labor intensive.

Over the last several hundred years, the cohesive matrilineal corporate groups that characterize Manjako social organization have ensured rotating land tenure and the collective labor necessary for successful rice farming in a cashless economy. The Manjako of Caio belong to five matrilineal clans, each of which has three to ten subclans. Each subclan has a residential court,[2] which controls rice fields and also large areas of mangrove, which is harvested for firewood and seafood. When the headman of a matrilineal court dies, his successor is chosen from among his matrilineal (male) relatives, usually a younger brother (of the same mother) or a maternal nephew. The residential system is semi-harmonic (Dugast 1995, 122); while the Manjako of Caio are matrilineal, residence is patrilocal and virilocal: men live in their fathers' courts unless and until they inherit the headmanship of another matrilineal court. So a court is made up of the current headman and his descendants, as well as the children, grandchildren, and great-grandchildren of past headmen. All dependents of a court—whether male or female, whether residents or not, whether belonging to the subclan or not—are entitled to the use of the court's rice fields; exactly who gets which parcel is ultimately decided by the current headman.

Outer dikes that protect the rice fields from seawater are in constant need of upkeep, and new dikes must occasionally be built. Dike work is organized by headmen who have the authority to call on their male dependents for such labor (Fig. 2.2). Different sections of outer dikes are under the responsibility of different headmen. If one headman lets his section of the dike erode, an entire rice field with sections belonging to several courts can be flooded and ruined.

The division of labor for rice work is along gender lines and relatively even; the work is almost always collective. Men, besides building and maintaining dikes, also plow the rice nurseries and the fields using a specialized, long-handled fulcrum shovel with a flat, narrow blade to turn the soil into furrows. Women transplant the rice seedlings from the nursery to the fields, weed, harvest, thresh, and carry the rice to huge granaries where it is stored for year-round consumption (Fig. 2.3). Women's work may be less intensive but it is long and exhausting, especially under the hot sun. Women and girls may walk miles to and from the rice field almost daily; at harvest time, they are loaded down with food and water on the way out and huge bags of rice on the way home.

Fig. 2.2 Men building new dike

Fig. 2.3 Rice granaries

Both men and women have rice granaries. The granary of the male household head is the largest, since women usually contribute to their husband's or father's granary; his granary is generally used to feed the household. But women, including widows, also have their own stores that can be quite large and that they use for their own children and special events. Rice can also be "traded for salt, storage baskets, peanuts, potatoes, manioc and fish" (Constantine 2005, 31).

Besides rice, people have long cultivated other foods in upland fields or gardens bordering the wet rice fields, including upland rice, sweet potatoes, peanuts, beans, okra, squash, tomatoes, and several kinds of greens. Several kinds of fruit are grown locally, including banana, mango, papaya, orange, and lime; there are also several species of native trees that produce edible fruit, such as baobab. Oil palms provide oil for cooking, sauces, and lamps (another labor-intensive task for women). Villagers also keep livestock including cows, pigs, goats, chickens, ducks, and guinea fowl; all but the latter two are generally used only for ceremonial purposes. However, ceremonies are frequent, so eating meat is not a rare occurrence. Women and youngsters also harvest clams, oysters, shellfish, and small fish from the mangrove. Small game is still hunted locally, though it has become very sparse, and honey is sometimes collected. Village diet is still so good that "students [sent to live with extended family in Bissau] say life is better in the rural areas, where a person eats three meals per day as opposed to one and where seasonal food is freely exchanged" (Constantine 2005, 117).

Even in the 1990s, villages on the outskirts of Caio still had a virtually cashless economy, except for a few items sold in stores, such as candles, batteries, matches, laundry soap. People produced a variety of food, raised livestock, collected fish and shellfish in the mangrove, wild fruits, small game, and honey. They built houses, furniture, and household items out of local wood and plant fiber. They made their own soap and salt, and made lamps out of palm oil. Potters still make cooking pots, and women weave mats and baskets. Older women still know how to spin cotton into thread for weaving.

Food, homemade household items (such as brooms, seats, mats), and services were—and in some places still are—not for sale, though they can be bartered or traded. Nor were—are—animals, which are always reserved for a specific ritual (see Journet-Diallo 2007; Constantine 2005). This tradition continues to a large extent, and in many villages, it is still difficult to buy cows, pigs, or chickens, though they are present everywhere.

Two local indigenous products, however, have long been sold as commodities: palm wine, sold fresh locally, and palm oil, sold in market towns. Palm wine is harvested from palm trees individually owned by men and passed down to their sons or other male relatives. Except during the rainy season, lads scale the palms twice a day and collect the cool, fresh, fermented wine in recycled liquor bottles in the early morning; it is usually distributed or sold by mid-morning (it goes flat quickly and doesn't store well). Women often act as middlemen, keeping track of orders and sales. Because women make their own palm oil, it is rarely sold locally, but instead, surpluses are sold in market towns far from the coastal areas where palm-oil palms are abundant.[3]

Cashew Production

When I first arrived in Caio in 1991, though there were already many large cashew orchards, most of the land between the villages and the rice fields and between villages was bush. But since then, virtually all land between and around residential compounds has been planted with cashews. Land that had been used to grow food or was a habitat for native plant and animal species has now been turned into cashew orchards.

Local people still remember government agents coming to the Caio with cashew seeds. One man said that cashews were first planted when, after independence, the government, through village committees, gave cashew seeds to people to plant. After that, it was up to each grower to clear more fields and plant more trees, until finally *mkaju a cas breng*, "cashews have chased away the bush." Another man agreed and added that the state also gave out peanuts. After the first year, it was up to each individual to continue growing the cashews or peanuts. He said, as others have, that cashews help with expenses when there isn't enough rice.

Cashew trees start producing fruit after three years; full production is reached by the tenth year. Trees continue to bear until they're about

30 years old. They produce soft, juicy apples ranging in color from pale yellow to dark red. The edible kernel grows inside a hard shell and hangs from the apple (Fig. 2.4). The shells are very hard and difficult to remove from the nut meat. Just inside the outer shell is a coating of resin that is toxic and spatters when heated. Cashew kernels can be removed from the shells locally, but it is a very painstaking and dangerous process. The nuts still in their shells are heated on a sheet of metal over an open fire until they burn at very hot temperatures, spitting boiling resin and spewing noxious fumes; then, the shells are cracked open one by one with a rock or hammer, and the kernel is removed. Because of the difficulty involved in hulling the nut safely, the huge majority of nuts are exported for industrial processing.

Women, after separating the nut from the apple in the orchard, carry the nuts to the homesteads and pour them into 50-kilo sacks. The bags of nuts are exchanged for imported rice and/or for cash. Flatbed trucks arrive at each homestead for the exchange; nuts are then transported to the capital, where they are exported for shelling (there is a single cashew processing station in the country, in Quinhamel, and it is not always in working condition). The exchange rate varies depending on the market prices of cashews and of rice and on the phase of the cashew season.

Fig. 2.4 Cashew apples

In the years I was present, people traded 2 kilos of rice per kilo of cashews at the beginning of the season. In the middle of the season, it was 1 kilo of rice per kilo of cashews. By the end of the season, it was 1 kilo of rice for 2 kilos of cashews. In 2013, it was 1 kilo of rice for 3 of cashews.

The juice is processed locally. Women press the fruit in "canoas"—literally, canoes—with a hole at one end for the juice to funnel out into containers (Fig. 2.5). The juice is then left to ferment to make "cashew wine" or is distilled into alcohol referred to locally as *cana*. Cashew wine is not appropriate for ceremonies at spirit or ancestor shrines: "Newer products like cashew wine, red wine, and beer are unsuitable for this purpose" (Constantine 2005, 42). However, *cana* is used at ceremonies, since it is a hard liquor equivalent to rum. Havik (1995, 31–32) has described how *cana* is also used to pay work groups to work in the rice fields.

The division of labor for cashew work is lopsided. Men plant—and therefore own—the trees. Once the trees are planted, there is little for the men to do except occasional pruning. On the other hand, during cashew harvest (the "campanha"), women and girls spend long hours gathering and processing cashew apples. It's a second harvest season in March–May, similar to the rice harvest in November–December. Though not as physically demanding as threshing, winnowing, and carrying rice, harvesting cashews includes ceaseless bending to pick up individual cashew apples from the ground, separating the nuts from the apples, and then crushing the apples to make juice. And women are already overworked: "In a rough survey of how women spent their day in the area it was found that a minimum of six hours was spent in housework (mainly meal preparation), while another six hours were spent in the fields. Women stated that they had no leisure time and very little time for any other activities (direct observation by J. Ki-Zerbo and R. Galli, November 1984)" (Galli and Jones 1987, 144).

In recent years, for reasons that will be discussed below, many families run out of rice before the end of the year and use their profits from cashews to buy rice from Asia. So, basically, instead of growing their own rice (and they used to have a surplus), people now grow cashews in exchange for rice from Thailand.

Fig. 2.5 Women juicing cashew apples

Trade-Offs

Each of these two economic activities—rice farming and cashew farming—has its own foundations, entailments, and consequences. Rice cultivation is both land intensive and labor intensive. Rice fields don't belong to individuals but to corporate matrilineages. The division of labor is fairly even: men build and maintain dikes, prepare the nurseries, and plow the rice fields. Women transplant the seedlings from the nurseries to the fields, harvest, thresh, winnow, and carry the grain to the granaries. Rice is eaten in family groups. Though women contribute rice to the granaries of their male relatives, they also have their own rows of rice paddies, and rice from those paddies goes into the women's personal granaries. Rice farming, supplemented by seafood and other cultivated crops (such as peanuts, squash, and beans), offers food security. Headmen, because they control the rice fields on which everyone depends, have the authority needed to maintain social stability.

Cashew trees are individually owned. The division of labor is uneven: men clear land and plant the trees; women collect the fallen apples, separate the nuts from the fruit, and carry the nuts to the house. Women also juice the apples to make cashew wine. Cashew money goes the "owner" of the tree (usually a man), though he may share it with the women who do the work and use it on his family. Money is individual, goes into pockets, and is used to purchase food and other commodities. Because wealth now comes from cashews, young men no longer depend on elders for access to rice fields, elders' authority has diminished, and some are unable to command the necessary labor to keep up the rice fields—in particular, the dikes. Since the dikes are no longer well maintained and rice fields may not be plowed, women cannot fill their granaries and are increasingly dependent on men and on cashew profits. Rather than a diet of locally grown food, people now eat a lot of processed, store-bought food; children now have lots of cavities, and the chronic diseases of developed countries (diabetes, hypertension, obesity) loom. The cashew economy brings a high risk of food insecurity.

The differences between rice and cashew farming are summarized in Table 2.1.

Cashew farming does offer several advantages. Cashew trees are said to grow in sandy soil where other trees will not. Cashew production is not as labor intensive as rice farming—especially for the men, who, after planting the trees, have virtually no work to do. Most importantly, though, it brings in cash from at least three sources: the sale of the nuts, the sale of cashew wine and *cana*, and the wages for young men, who are hired by cashew traders to load and unload 50-kilo sacks of cashews and rice onto and off of the large flatbed trucks that reach even far-flung homesteads. Furthermore, enterprising local men who manage to put

Table 2.1 Rice vs. cashews

	Rice	Cashews
Control of resources	Collective	Private
Gender division of labor	Fairly even	Lopsided
Product	Eaten, distributed	Sold, traded
Lineage authority	Strong	Weak
Women's economic status	High	Low
Diet	Good	Poor
Food security	High	Low
Environment	Friendly	Harmful

together enough capital can invest in a truck and become traders themselves. The extra cash from cashews is used to purchase common store-bought items such as batteries, cloth, and soap. Temudo and Abrantes relate that cashew farmers in other regions of the country "were able to buy tin roofs for their houses (a major goal for farmers), bicycles and motorcycles, and even to discontinue or drastically reduce cereal production and, in the case of Fula and Mandinga farmers, to increase the use of wage labor (mainly with migrant workers from neighboring Guinea)" (Temudo and Abrantes 2014, 226). And, as was stated earlier, the revenue from cashew exports on the national level finances the state government.

Cashews may be particularly advantageous to women, especially as their economic cornerstone—rice—becomes unreliable due to dilapidated dikes and abandoned rice fields. Because cashew orchards are individually owned, women can inherit them from their fathers, brothers, or husbands; they then keep all profits from the sale of nuts for themselves. Cashew apple "wine" and cashew *cana* are women's home industries from which they keep the profit. Havik (1995), citing Cornelius Hugo and Cardoso (1990) and van der Drift (1990), emphasizes the additional economic power the sale of *cana* brings women, who can also use the profits to hire male work teams to maintain rice fields.

On the other hand, cashews bring many kinds of problems, ranging from economic to environmental to social. Monoculture is inherently risky: a disease could wipe out the cashew crop and would leave the country dependent on food handouts. The life span of the trees is

50–60 years, 30 of which are fruit-bearing. Trees planted in the 80–90s will soon stop bearing fruit, and the land will be covered with dead, rotting trees—in fact, in some orchards it already is, both in Caio and elsewhere (see Temudo and Abrantes 2014). Increased labor and investment will be needed to replant the orchards. Horticultural plots that were rotated and lay fallow are gone.

Environmental threats are real. As already stated, cashew trees grow well in sandy soil; however, they deplete that soil. Local people have already noted, with apprehension, that little grows under the cashew trees. Biodiversity has disappeared. Bush is gone. Medicinal plants are disappearing. Animal species are disappearing; for example, in the 1990s I often saw five-foot tall marabout storks in the bush near the village, but nevermore. Hunting is virtually nonexistent. Many of the non-governmental organizations (NGOs) and World Bank agencies are encouraging the use of pesticides to increase the yield, with unknown effects. Non-biodegradable trash is accumulating, with no slowing in sight. Cheap plastic and paper products are now being bought and trashed at astonishing rates. Used and broken items are tossed anywhere. There is no traditional way of dealing with trash because there never used to be any; now it's a problem.

The cash economy brings about wealth disparity, haves vs. have-nots. The disparity causes envy, jealousy, and ill will, which lead to accusations of using spiritual means to harm others. Such feelings came to a head in 2004, and the women of Caio brought a new spirit shrine specifically to counteract the alleged ill will of men toward the welfare of women and children (children belong to the mother's lineage, not the father's, so a man's children are not his heirs). Furthermore, with the growing influx of cash, the costs of ceremonies have skyrocketed. Rather than a single cow for planting an ancestor post, three or four cows must be killed, not to mention enormous amounts of alcohol and food. Much of the cost of ceremonies is paid for with cashew money, but returning migrants are also hard hit. If they don't pay tribute to local spirits and ancestors, they and their families are at risk of illness and death. Odile Journet-Diallo (2007) gives a detailed account of inflation in ceremonial rituals among the Diola, a neighboring and closely related ethnic group. Finally, Lundy (2012) has pointed out an increase in alcoholism corresponding to the increased availability of cashew wine and *cana*.

Turnaround?

In the last few years, I've noticed a shift occurring in many households. Discourse is changing. People are saying the city isn't such a good place to live and that they eat better in the village. They've realized that they can't count on high prices for cashews and low prices for imported rice. They also miss the variety of foods they once had.

Rather than vocally contesting the growing dependence on cashews vs. the traditional dependence on rice, they are "voting with their hoes." In 2015, I discovered several acres of bush had been cleared in order to grow upland rice, which used to be grown in several areas of Caio before cashew orchards spread across the land. Upland rice, which depends solely on rainwater, is a lot of work and doesn't produce as much as paddy rice, but it doesn't require dikes so can be grown by small family groups. Everyone clears the bush; men turn the soil; women sow, weed, and harvest. A young man I met at the new field explained that they'd cleared the whole section that year for the first time since 1996. They'd planted upland rice, but he said that the next year they might plant peanuts.

The next year they planted a lot more than rice and peanuts. In 2017, women in one household proudly showed me the crops they'd grown in the new upland field the previous year: rice, peanuts, corn, sweet potatoes, cucumbers, okra, beans (*u-fó*), and roselle (*Hibiscus sabdariffa* L.). They ground the corn into a fine meal to make porridge for young children; the rest of the crops were eaten by the family or, if there was a surplus, traded or sold. They showed me the seeds of these plants they'd saved to plant the following year. In fact, though male relatives always turn the soil, the clearing and cultivation of the upland field seem to have been directed by enterprising women ranging in age from 30 to 60 years.

Also in 2017, I discovered that a dike was being built far out into the mangroves in order to recuperate rice fields that had long been abandoned and had reverted back to impenetrable mangrove. A team of men from one of the local homesteads, under the leadership of the household head, were putting in long hours of sweaty, muddy, backbreaking work (Fig. 2.2) to rebuild the dike where it had once stood and thus recuperate rice fields belonging to the matrilineal court. The investment in time and labor would be enormous; once the dike is finished, the mangrove itself will be cleared and then turned into furrowed paddies. The dike will keep out the seawater and hold in rainwater, so that after five or six years, the field will be ready for planting. So, within spitting distance of abandoned rice fields, at least one old rice field is being recuperated.

Another optimistic sign of economic and social perseverance is that young men who went to Bissau or Senegal seeking money-earning jobs have returned with wives and are joining in the rice work and raising families.

Broader Reflections

In addition to showing some of the effects of cashew production and how people in Caio are faring, this case study invites broader reflections on several questions. Who profits from "development"? Does growing cash crops necessarily replace traditional farming, or can people continue to do both? Finally, why do some households and villages farm rice successfully while others don't?

There are many places in the world where people produce enough food to sustain themselves quite well yet they do not export a surplus or a cash crop. So who profits from an export economy? Walter Rodney (1973) was one of the first to say that it was not small-scale farmers, but rather colonial governments. Yet even today, state governments continue to push economic development—usually in the form of cash crops—in order to support themselves.

Many authors writing specifically about Guinea-Bissau have agreed. For example, "an increase in agricultural production is necessary to help feed an urban workforce, provide commodities to export in exchange for "hard" currency, supply raw materials for industry, and finance the institutions and personnel of the government" (Galli and Jones 1987, 187).

From the beginning, the goal was to amass revenues to support the state. "The search for revenues to finance the state apparatus has been the dominant economic concern of state policy" (ibid., 189).

Forrest goes further, proposing that a strong state government and strong, self-reliant villages are at odds: "The combination of a strong rural civil society, a national administration that was politically fragile and disconnected from the major social formations in the rural areas, and a vibrant informal economic sector all diminished the ability of the state to consolidate its power, to tax its populace, and to implement macro level policies on a national basis" (Forrest 2003, 227). In other words, Forrest is saying that the state is weak *because* villages are strong. Perhaps not coincidentally, at the European Conference on African Studies (ECAS) 2014 conference in Paris, one panel (215) was titled "Economic failure, political success?" Perhaps the question should be reversed: economic success, political failure?

While the World Bank, USAID, and most other agencies continue to push cashew production, at least some researchers have pointed out the danger of overdependence on cashews. For example, "Cashews represent 90 percent of the country's exports and the principal source of income in rural areas.… [O]ver the last three decades, the production of rice has significantly decreased in favor of cashew farming. This situation represents a threat to food security" (Barry et al. 2007, 77). Unfortunately, this warning seems to be going unheeded as cashew production continues to increase.

The most commonly stated goal of development programs is to reduce poverty. But how poverty is defined and measured is problematic. Most scales are based on GDP or on per capita GDP. But these figures are based only on countable consumption and exports, not on food grown locally that stays local, that goes from field to kitchen, or from family to family, without being weighed or measured. A local rice farmer told me in 2016, "we have enough rice for three years," not "we have 250 kilos of rice" or "we have $300 worth of rice." Because economists necessarily count only countables (tons, gallons, dollars), people in subsistence or informal economies—in which cash may not play a role—are often labeled as poor. For example, "People are poor because they lack productivity. They lack productivity because they don't have the tools to become more productive. Those tools include the basic inputs to raise farm yields above subsistence levels" (Shinkle 2008). Yet Caio rice farmers are among the most productive in the world. They have a sustainable, environmentally friendly economic system that offers food security. They also have an incredibly tight-knit social system, upheld by their religious practices, that leaves no one out. This situation exemplifies Hobart's observation that, in Western discourse, there is often "the presumption of the priority of technology over social considerations" (Hobart 1993, 3).

In many areas of Guinea-Bissau, including Caio, rice production is reportedly diminishing (see Barry et al., op. cit.). "Even those households [in Caio] for whom wetland cultivation was their most productive source of rice did not have enough for the coming year. […] Everyone supplemented his or her rice store through the cashew harvest, traded some or all of the nuts for imported rice, using this before opening stores of cultivated rice" (Constantine 2005, 192). Scholars have suggested various reasons for diminished rice production. For a neighboring Manjako kingdom, Gable (1990) emphasized male outmigration, which has resulted in a labor shortage; in Caio, a high rate of male outmigration also likely contributes to falling rice production. Davidson (2012, 18) for her part attributes insufficient rice among the Diola (rice-growers just north of the Manjako) to climate change, including declining rainfall and desertification.

However, although rainfall does vary by year and by decade, data show fairly consistent precipitation levels in the area. A dataset produced by the Climatic Research Unit (CRU) of University of East Anglia (UEA) published on the World Bank website shows that since several relatively dry years in the 1980, rainfall averages are back within the normal range, with year-to-year variation. Moreover, Bacci, Diop, and Pasqui state that for the middle and upper Casamance (in southern Senegal, just north of Guinea-Bissau), there has been "a tendency over the last years of *an increase* in temperature, in potential evapotranspiration, and *in precipitation*" (Bacci et al. 2013, 12–13, my translation and my emphasis). In Caio, some people say there's not enough rain while others say there's plenty. Lack of rainfall would affect all households equally, yet some households have had bumper rice crops in recent years.

So the question remains as to why some villages in the rice belt—and some households within those villages—are not as productive as others. In Caio, where Constantine observed a lack of rice, in many households rice is plentiful. Villages are often seen as homogenous; yet each family, each compound is made up of members of different personalities, networks, ambitions, values, skills, and risk tolerance. For some, the effort of clearing mangrove is worth the food security; for others, cash from cashews is the way forward. In some households, a single enterprising individual will clear and fence a new garden plot. Some charismatic individuals—especially in strong matrilineal courts—entice others to join him or her in new endeavors, such as clearing a new upland field or rebuilding dikes to recuperate rice paddies. In other villages and households, especially when men have emigrated in high numbers (see Gable 1990), it is less likely that families can or will maintain their rice fields and resist the trend to depend on cashew profits for their livelihood.

Can Caio residents grow both rice and cashews? The answer could be "yes." Rice and cashews do not compete for land. The Manjako have a figurative and literal attachment to the land, which binds people together. Rice fields cannot be bought and sold; they are in the hands of corporate groups, so rice paddies will always be available for individuals based on their kin group and compound of birth. Cashews, on the other hand, grow in upland areas that used to be unclaimed bush. The skills and knowledge necessary to grow rice are being transmitted; children learn by doing from a very young age. There is a very low level of state interaction, which could augur well for village economic independence. Finally, many people—especially women—are using cash from cashews (nuts, wine, and *cana*) not only to hire rice field laborers but also to dehusk rice by machine rather than by pounding it with mortar and pestle. In other words, they are using cashew money to increase their rice production.

On the other hand, the rise of cashew farming has indeed coincided with reduced rice output, likely because of shortage of labor. The kind of rice farming practiced in coastal Guinea-Bissau requires intensive, collective labor, particularly by the men to build and maintain dikes and plow the fields. But men now have other options, so dikes are falling into ruin and entire fields are being flooded. When one family's dikes fail, other families are ruined. Women are dependent on men for the upkeep of the dikes, preparing and sowing the nurseries, and plowing the paddies in preparation for transplanting the rice. But this is still a situation of selling a cash crop to acquire more rice, which has historically been produced in huge surpluses. Moreover, women already have the annual rice season and harvest to deal with, on top of housework and childcare; as noted above, they have very little free time. A second, cashew harvest is physically and socially overwhelming.

Villagers are often said to be mired in traditions and cultures that keep them from "developing." "Whatever the rationale, non-western societies have been widely represented as static, passive and incapable of the progress based on rational government and economic activity which the West alone could provide" (Hobart 1993, 2). Like farmers in Mali described by Wooten, Caio residents, rather than being stuck in a monolithic traditional past, "are actually forward-looking individuals, individuals who actively negotiate new circumstances, improvising solutions to new challenges by drawing on and restructuring or rejecting established social, cultural, and economic patterns" (Wooten 1997, xvii). They are weighing their options, taking advantage of opportunities, old and new, and deciding their own economic activities.

All over the world, so-called developed countries have sacrificed biodiversity for "amber waves of grain"; forests and prairies have been cleared and plowed at the expense of wildlife and clean air and water. It could seem hypocritical to sound the alarm that the same is now happening in Africa. But much is at stake, especially if the reason for the destruction of natural bush is revenues for state cadres and profits for international traders. State and international agencies that encourage cash crops at the expense of sustainable subsistence activities should reconsider their "development" programs, especially when, as noted above, they lead to both food insecurity and threat to the environment. Thankfully, in Caio, as elsewhere, many have realized that the old, reliable way can be better and that they can live off their own land through their own hard work.

Notes

1. I worked for the Medical Research Council from 1991 to 1993 on an interdisciplinary HIV/AIDS research project based in Caio and since then have returned to the village every two or three years for a period of one to five months of ethnographic research.
2. Besides matrilineal courts, there are also newer patrilineal homesteads governed by the eldest male resident; these homesteads also control rice fields that were granted to them by matrilineal headmen.
3. The kingdom is renowned for its palm wine. Its very name, Caio, comes from *kayu*, to pierce, referring to the way young men pierce the palm to harvest the wine.

References

Bacci, M., Diop, M., & Pasqui, M. (2013). *Encadrement climatique et évaluation du changement climatique dans les régions d'étude* (Rapport No. 6, pp. 12–13). Dakar: Program d'Appui au Programme National d'Investissement de l'Agriculture du Sénégal.

Barry, B. S., Creppy, E., & Wodon, Q. (2007). Cashew production, taxation, and poverty in Guinea-Bissau. In B. S. Barry, G. E. Creppy, E. Gacitua-Mario, & Q. Wodon (Eds.), *Conflict, livelihoods, and poverty in Guinea-Bissau* (World Bank Working Paper No. 88). Washington, DC: World Bank. https://openknowledge.worldbank.org/handle/10986/6879, License: CC BY 3.0 IGO.

Catarino, L., Menezes, Y., & Sardinha, R. (2015). Cashew cultivation in Guinea-Bissau—Risks and challenges of the success of a cash crop. *Scientia Agricola, 72*(5), 459–467. https://doi.org/10.1590/0103-9016-2014-0369.

Constantine, S. (2005). *Locating danger and negotiating risk on Manjaco Terrain for Bayu, the children of Caió, Guinea Bissau* (PhD dissertation, University of Cambridge, Cambridge, UK).

Cornelius, H., Lea, J. D., & Cardoso, C. (1990). *A review of the cashew sub-sector in Guinea Bissau*. Manhattan, KS: Kansas State University and Bissau: Instituto Nacional de Estudos e Pesquisas. pdf.usaid.gov/pdf_docs/pnabg194.pdf.

Davidson, J. (2012). Basket cases and breadbaskets: Sacred rice and agricultural development in postcolonial Africa. *Culture, Agriculture, Food and Environment, 34*(1), 15–32.

Dugast, S. (1995). Lignages, classes d'âge, village. A propos de quelques sociétés lagunaires de Côte d'Ivoire. *L'Homme, 134*(3), 111–157.

Forrest, J. B. (2003). *Lineages of state fragility: Rural civil society in Guinea-Bissau*. Athens: Ohio University Press and Oxford: James Currey.

Gable, E. (1990). *Modern Manjaco: The ethos of power in a West African society* (PhD dissertation, Department of Anthropology, University of Virginia, Charlottesville).

Galli, R. E., & Jones, J. (1987). *Guinea-Bissau: Politics, economics and society*. London: Frances Pinter; Boulder, CO: Lynne Rienner.

Havik, P. (1995). Relações de género e comércio: estratégias inovadores de mulheres na Guiné-Bissau. *Soronda* (Bissau), *19*, 25–36.

Hobart, M. (1993). Introduction. In M. Hobart (Ed.), *An anthropological critique of development: The growth of ignorance* (pp. 1–30). New York: Routledge.

Journet-Diallo, O. (2007). *Les Créances de la Terre: Chroniques du Pays Jamaat (Jóola de Guinée-Bissau)*. Turnhout, Belgium: Brepols.

Linares, O. (1992). *Power, prayer, and production: The Jola of Casamance, Senegal*. Cambridge and New York: Cambridge University Press.

Lundy, B. D. (2012). Playing the market: How the cashew 'commodity scape' is redefining Guinea-Bissau's countryside. *Culture, Agriculture, Food, and Environment, 34*(1), 33–52.

Pélissier, P. (1966). *Les Paysans du Sénégal*. Saint-Irieix, France: Imprimerie Fabrègue.

Rodney, W. (1973). *How Europe underdeveloped Africa*. London: Bogle-L'Ouverture Publications and Dar-es-Salaam: Tanzanian Publishing House.

Shinkle, K. (2008, April 21). Where markets don't work (review of Jeffrey Sachs' book *Common wealth: Economics for a crowded planet*). U.S. News and World Report, p. 78.

Temudo, M. P., & Abrantes, M. (2014). The cashew frontier in Guinea-Bissau, West Africa: Changing landscapes and livelihoods. *Human Ecology, 42*, 217–230.

UNIOGBIS (United Nations Integrated Peacebuilding Office in Guinea-Bissau). (2016). *Cashew nut central to Guinea-Bissau economy: A blessing or a curse?* http://uniogbis.unmissions.org/en/cashew-nut-central-guinea-bissau-economy-blessing-or-curse. Accessed April 7, 2016.

USAID (United States Agency for International Development). (2005). *USAID joins private sector to boost Guinea-Bissau cashew revenues*. http://senegal.usaid.gov/news/releases/05_06_28_GB_cashews.html. Accessed February 4, 2008.

Van der Drift, R. (1990). O desenvolvimento da produção e to consumo de álcool entre os Balanta Brassa da aldeia de Foia, no sul da Guiné-Bissau. *Soronda* (Bissau) *9*, 95–116.

Wooten, S. (1997). *'Gardens are for cash, grain is for life': The social organization of parallel production processes in a rural Bamana village (Mali)* (Doctoral thesis, University of Illinois at Urbana-Champaign, Urbana-Champaign).

3

Trade-Offs Between Crop Production and Other Benefits Derived from Wetland Areas: Short-Term Gain Versus Long-Term Livelihood Options in Ombeyi Watershed, Kenya

Serena A. A. Nasongo, Charlotte de Fraiture and J. B. Okeyo-Owuor

Introduction

Through absorbing and processing waste products, wetlands play a critical role in sustaining the value of the environment. 'Wetlands' are commonly defined as 'natural areas where water meets land,' or more specifically as '…lands that are saturated with water long enough to cause

---

S. A. A. Nasongo (✉)
University of Amsterdam, Amsterdam, The Netherlands

C. de Fraiture
Land and Water Development, IHE Delft, Delft, The Netherlands
e-mail: defraiture@unesco-ihe.org

J. B. Okeyo-Owuor
Rongo University College, Rongo, Kenya

the formation of waterlogged (hydric) soils and the growth of water-loving (hydrophytic) or water tolerant plants.'[1] They are one of the most complex and threatened types of landscape in the world and almost universally under threat. By biologically cycling carbon dioxide, methane, and hydrogen sulfide, they sequester (trap) and release carbon, thus regulating climate change effects. Wetlands also sustain a rich multiplicity of plants and animals which help to preserve wetland processes such as nutrient cycling, sediment trapping, and water storage. The loss of wetlands universally is a well-known problem and is detrimental, but continues, with little public recognition of the causes or consequences. Some causes of degradation include erosion, salinization, pollution, invasive species, loss of vegetation, eutrophication, sediment deposition, and different hydrological regimes occurring as a result of ever-increasing agriculture and development (Davis and Froend 1999; Gren et al. 1994).

The occupations of people living within or on the periphery of wetlands often depend partly or entirely on wetland 'ecosystem services,' i.e., profits that can be drawn from their use or exploitation. Loss or degradation harms them directly. 'Ecosystem service *trade-offs*'—central in this chapter—arise from the reduction in one ecosystem service as a result of an increase in the delivery of another ecosystem service. Such trade-offs may be the outcome of economic or natural processes or tough management choices (Cohen-Shacham et al. 2011; Millennium Ecosystem Assessment (MEA) 2005a).

All over the world, large tracts of land supporting a broad range of activities including farming and irrigation are commonly connected with wetlands. While wetland agriculture can bring substantial benefits regarding food security, health, and income, unwise development often results in wetland degradation, detrimental environmental impacts, and damaging consequences to people's livelihoods. These agricultural activities modify the hydrological regimes and other natural aspects on which wetlands depend, and hence exert a broad range of mostly adverse impacts on their ecological character and the other benefits they offer (McCartney et al. 2010). Declines in the delivery of services such as freshwater and some fish species result from the extensive use of water for irrigation and excessive nutrient loading accompanying the use of nitrogen and phosphorus in fertilizers (MEA 2005b).

Irrigation schemes or water diversions for irrigation have thus undeniably caused adverse effects to wetland ecosystems. One effect of irrigation is the replacement of wetland vegetation by highland flora which consequently affects the biota that depends on these wetlands and the services that people obtain from these ecosystems (Galbraith et al. 2005). Through better control of water, farmers can prolong the cropping period and reduce threats arising from the effects of either drought or flooding. However, there is often little thought of the broader environmental impacts and the consequences for other ecosystem services and livelihoods based on the provisioning services of wetlands (McCartney et al. 2010).

Agriculture within and on the periphery of wetlands can lead to conflict between farmers and other wetland users. The most common impact of the development of market–orientated wetland agriculture is the loss of subsistence farming, which is often associated with mixed cropping and less water use. Market–orientated wetland agriculture is based on monoculture and intensive water use. Agricultural intensification in wetlands, therefore, results in groups of people dependent on subsistence farming losing out, with a negative response cycle occurring where productivity losses lead to further expansion and transformation of wetland areas (Wood and van Halsema 2008).

A range of hard choices, based on a sound ecological, social, and economic analysis, will be needed for wetland conservation since future development investment must be grounded on more careful consideration of the long-term costs of converting natural systems, for example, conversion of natural wetlands to intensive agricultural production or reclaiming wetlands to provide space for industrial or urban expansion (Wanzie 2003). The MEA defines ecosystem services as 'The benefits provided by ecosystems to humans, which contribute to making human life both possible and worth living' (MEA 2005b, 23). Furthermore, the MEA outlines specific classes of ecosystem services, including:

- *Provisioning services* (e.g., material goods such as food, feed, fuel, and fiber);
- *Regulating services* (e.g., climate regulation, flood control, water purification);

- *Cultural services* (e.g., recreational, spiritual, aesthetic); and
- *Supporting services* (e.g., nutrient cycling, primary production, soil formation).

Conservation and development need not be conflicting goals but are better considered as part of a political and social process of engagement and compromise among diverse interests (Brechin et al. 2002). The link between irrigated agriculture and its effects on wetland systems has frequently been shown as one of a direct trade-off between the human need for food versus the environment (Galbraith et al. 2005). Trade-offs are multifaceted, appear in many semblances, and are looked at differently by various people, depending on their interests (Sunderland et al. 2008). The aim of addressing trade-offs between cropping wetlands and harvesting wetland products should ideally not be concentrating on conservation and poverty reduction, but rather to enhance the net benefits for people while at the same time avoiding major ecological threats and ensuring the long-term sustainability of different ecosystem services (Senaratna Sellamuttu et al. 2008). For such trade-offs to be acceptable and viable, the options offered have to offset or balance the losses caused by abiding with the restrictions and these choices can be in monetary terms, employment, or access to other resources (Schmidt-Soltau 2004). Trade-offs under consideration among various wetland ecosystem services need cooperation across sectors and the different stakeholders who are instrumental in planning activities in support of sustainable use of these delicate ecosystems. Weak regulation and accountability systems and sectoral-based institutions constitute a significant obstacle to effective management of wetland ecosystems. Trade-offs between and within the objectives of conservation and development are the rule rather than the exception, as the creation of protected areas always entails trading one land-use option for another, and choices among different interests have to be constantly faced.

Development policies have often neglected the small-farm and resource users whose rising disposable income could drive development. To correct this anomaly, choices have to be made which include: the participation of the local communities at all stages of wetland conservation and development, the development of policies for wetland conservation that contribute to sustainable growth, boosting small-scale development methods based on the efficiency of the natural ecosystem, and equitable sharing of benefits accruing from wetlands preservation and development (Wanzie 2003).

Inclusiveness is a fundamental normative rule of the ecosystem approach that emphasizes involvement of all stakeholders and balancing the local interests with that of the broader public. Inclusiveness requires meeting with a broad range of institutions, organizations, groups, and individuals that have an interest in, understanding of, or prospective influence over, the management of a given ecosystem. Participation should start at an early stage to clarify both the issues to be addressed by the decision-making process and where the priorities of stakeholders lie and also to identify constraining factors. There are many works on the challenges associated with stakeholder participation in natural resource management (Reed 2008). Localized management of the ecosystem increases the responsibility, ownership, accountability, participation, and use of local knowledge. Decision-making at the local scale is holistic and can accommodate the consideration of multiple benefits, trade-offs among benefits of ecosystem services, environmental limits, and proper levels of stakeholder participation. Where communities are furnished with appropriate information on the consequences of decisions, participatory approaches can improve outcomes. They provide new ideas to the process and facilitate discussion and analysis of the reasoning and assumptions behind decisions, but are unlikely to resolve fundamental conflicts in the management of natural resources on their own.

Our study aimed to use stakeholder analysis to enable a better understanding of the trade-offs between crop production, ecological benefits, and livelihoods derived from Ombeyi wetland. Specific objectives were: (1) to describe the wetland resources in Ombeyi wetland used by local communities; (2) to find out the perceptions of stakeholders on the situation of the swamp; and (3) to identify potential trade-offs between the various uses of the wetlands. All these are discussed in the perspective of sustainable use of the Ombeyi wetland and from there come up with recommendations for policy. This chapter describes the case of the Ombeyi catchment where recent trends and changes in livelihoods strategies put the still existing wetland areas under pressure.

Description of the Ombeyi Watershed

The Ombeyi wetland ecosystem is situated in Kisumu County, Kenya, and lies across two sub-counties, namely Nyando and Kisumu East. Administratively, the wetland ecosystem is found within Ombeyi location, which has five sub-locations, namely Obumba, Ramula, Kore, Ahero irrigation, and Kang'o, and North East Kano, which has six sub-locations, namely East Kabar, Central Kabar, West Kabar, Kamswa North, Sidho I, and Wang'aya II, respectively (Fig. 3.1).

The wetland ecosystem covers an estimated area of about 10.37 km^2 consisting of small swamps, streams, and rivers (covering 3.3% of the catchment area). The catchment area is approximately 318 km^2. The principal rivers and streams providing water to the wetland are Ombeyi, Oroba, and Nyakoko. The Oroba River originates from the Nandi Escarpment from numerous streams, like the Little Oroba and Great Oroba both of which later merge to form the Oroba, which enters the Ombeyi Swamp at Ombeyi. The Oroba leaves Ombeyi Swamp as the Ombeyi and Miriu. The Ombeyi wetland also receives

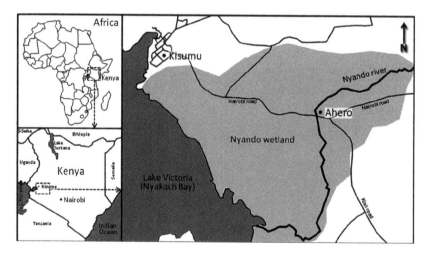

Fig. 3.1 Map showing the location of the Ombeyi wetland, Kenya (Coordinates: 0°11′–0°19′S, 34°47′–34°57′E) (*Source* van Dam et al. (2013). Linking hydrology, ecosystem function, and livelihood outcomes in African papyrus wetlands using a Bayesian network model. *Wetlands* 33)

water from rivulets such as the Nyangeto and Lielango which join to form the Mayenya. The climate in Ombeyi wetland is characterized by an annual rainfall averaging 1204 mm and distributed in three rainfall seasons March–May, August, and November–December. April and May comprise the wettest period of the year accounting for 25% of annual precipitation. The region falls within the Lake Victoria Lowlands and Kano Floodplains which are often flooded as they are surrounded by steep hills, specifically the Tinderet Hills to the east and the Nandi Escarpment to the north, and the Kano Plains which stretches from Miwani.

The Kano Plains is a flat lowland area with a topography that gently changes in the northeastern and southern parts. The highest point is found in Muhoroni at an altitude of 1801 m above sea level (asl) with the lowest point at the same altitude as the Lake (1134 m asl). In the Ombeyi watershed, large-scale sugar cane plantations, rainfed food crop cultivation, and rice production, as well as cattle grazing, are typical. On the farmed land, various crops are found, such as arrowroots, cassava, cowpeas, finger millet, green grams, groundnuts, kales, maize, onions, rice, sorghum, sugar cane, sweet potatoes, and tomatoes. In addition to these are varieties of local vegetables and fruits such as avocados, bananas, citrus, mangoes, and pawpaw.

Research Methods

A stakeholder analysis using the approach adapted from Darradi et al. (2006) was used (see Fig. 3.2) to identify potential trade-offs between the various uses of the wetlands, as shown in Table 3.9 in the Appendix. The methods employed to obtain primary data included a household survey in which 150 households were interviewed. The household survey was conducted between March 19 and 27, 2014 (see also Figs. 3.3 and 3.4). The sampling was done based on a stratified random sample. Five sub-locations were selected purposively to cover the upper, middle, and lower parts of the Ombeyi wetland, namely Central Kabar, East Kabar, Kore, Ramula, and Wang'aya II. A standardized questionnaire (composed of 40 questions) with open- and closed-ended questions

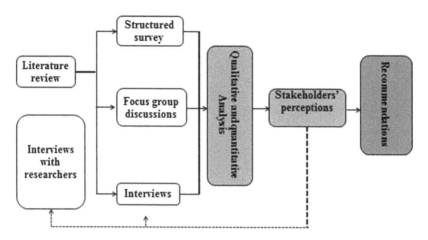

Fig. 3.2 General approach of stakeholder analysis (Adapted from Darradi et al. 2006)

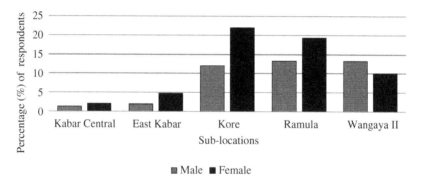

Fig. 3.3 Percentage of respondents by sub-location

was administered. The questionnaire had three principal parts, on (1) demographic information (age, gender, family size, and educational level), (2) economic activity (main activity, agriculture, livestock production, salt mining), and (3) stakes and perceptions of the users of the protected region and the effects of the current practices on their well-being. Participatory methods (community resource mapping, seasonal calendar, stakeholder identification, Venn diagram, conflict matrix, and problem ranking) and Key Informant Interviews (KII) were

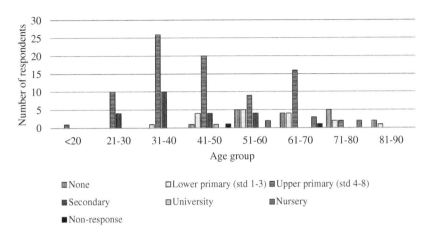

Fig. 3.4 Level of formal education of respondents by age group

also used (13 in total). The KII were conducted for government department, and line ministries relevant to the study, and these included agriculture, fisheries, water, physical planning parastatals like Kenya Wildlife Services (KWS), National Irrigation Board (NIB), Kenya Forest Services (KFS), KALRO (Kenya Agriculture and Livestock Research Organization); projects like Lake Victoria Environmental Management Programme II (LVEMP II); and authorities like National Environmental Management Authority (NEMA) and Water Resources Management Authority (WRMA). Community members were also used as key informants.

Fgures 3.3 and 3.4 present results from the household survey.

Results

Different Stakeholder Groups and Livelihood Strategies in the Ombeyi Watershed

After identification of the stakeholders within the Ombeyi wetlands, the stakeholders were categorized as government departments, civil society, faith-based organization, international organization, and others. Each of

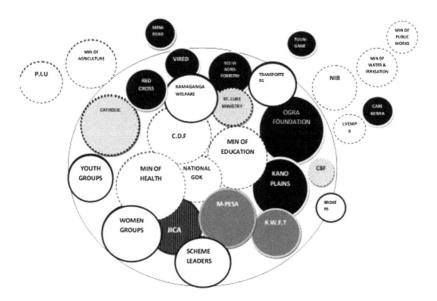

Fig. 3.5 Venn diagram for Kore sub-location

(*Key* white—community groups; black—CSOs; white with broken edge—government departments; gray with broken edge—faith-based organizations; gray—financial organizations; and striped black—donor)

these categories was given a different color as suggested by the participants in the FGDs. Using a Venn diagram (Fig. 3.5), the participants provided valuable insights into power structures and decision-making processes within the community. This method uses circles of different sizes to represent organizations. The bigger the circle, the more important is the institution to the community. The distance between circles represents the degree of influence or contact between agencies and the community.

Results from the FGDs show that for all the three sub-locations, institutions providing humanitarian and relief services to widows and orphans had the biggest circles. There were very few institutions involved in development and management irrigation and marketing of farm produce and also those dealing with environmental management and conservation. NIB and KALRO which are essential for crop irrigation and crop development, respectively, had a little contact with the community. Table 3.1 shows the relevant and potential stakeholders for sustainable use and management of the Ombeyi wetland.

Stakeholders

See Table 3.1.

Wealth Ranking and Water Use

A wealth ranking was done to determine the different wealth groups based on the assets they own regarding land and livestock. The very low ownership of livestock is immediately apparent; for example, even though the mean figures are not very low, the majority of households do not have any livestock (cattle, goat, or sheep), indicators of types of houses, utility people have access to, and types of their employment can be good proxies for the economic positions of households as well as their status in the community. To capture some dimensions of housing, whether houses have corrugated iron roof and whether the walls are built from bricks are reported. It was noted that most of the non-farm income is controlled by a relatively small number of households. The figures also highlight the relatively diversified nature of rural livelihoods; sources of income other than agriculture play important roles. Also, only the 1% of the families have access to piped water (Table 3.2).

Stakeholders' Livelihood Strategies

Stakeholders have a range of livelihood options. They diversify sources, depending on household characteristics, wealth, season, and others. The chief crops grown under irrigation within the smallholder farms popularly known as *achung' kenda* are arrowroot, maize, sorghum, sugar cane, and rice. Other crops grown include vegetables and watermelons. Some people derive their livelihoods from wetland products such as papyrus and fish. In most cases, wetland-based livelihoods include both crop production and harvesting of wetland products, depending on the

Table 3.1 Important and potentially interested parties in the Ombeyi wetland

Major and prospective stakeholders	Stake/Mandate	Interest	Pivotal role?
Farmers	Food security	Food security and improved incomes	Yes
Fishermen	Wetland resource users	Profit	Yes
Mat-makers	Wetland resource users	Profit	Yes
County government	Land ownership, Policy, and laws	Maintain law & order	Yes
National government (chiefs and assistant chiefs)	Maintain law & order	Security for development	Yes
NEMA	Conservation & management of natural resources	To safeguard & enhance the quality of the environment	Yes
WRMA	Continuous supply of water Control of soil erosion	Quantity & quality of water maintenance	Yes
Ministry of agriculture, fisheries, and livestock production	Enhance good agricultural practices	Food security and improved incomes	Yes
LBDA	Marketing of rice	Profit	Yes
KALRO	Research and crop development	Food security	Yes
LVEMP II	Conservation	Environmental protection	Yes
KFS	Environmental protection	Tree cover	Yes
Industries	Processing of produce	To make profit	Yes
Civil society organizations, e.g., Academy for Educational Development (AED)	Conservation & management of natural resources	Food security	Yes
NIB	Irrigation development	Food security and improved incomes	Yes
NOKISO	Marketing of produce	More food production	Yes

Table 3.2 Wealth categories in the Ombeyi wetland and water use

Wealth category	Description	How do they use water?	Percentage
Very wealthy	Scores of both big and small stock Tractor > 2 Permanent house (bricks/stone) Piped water Workers > 2 Land > 50 ac (sugar cane, rice) Businesses	Domestic use Irrigation Watering livestock Watering trees Washing machinery	1
Wealthy	Large and small stock Tractor-1 Permanent house (bricks/stone) Water from shallow well Workers-2 Land > 10 ac (sugar cane, rice) Posho mills Water pump for irrigation Fish farming Businesses	Domestic use Irrigation Watering livestock Water supply to community Watering trees Fish farming	20
Average	Cattle < 10 Goats 2 Sheep < 5 Semi-permanent house (mud wall plastered, iron roof) Land < 5 ac (arrowroots, rice, watermelon, sugar cane) Water pump Provide casual labor Harvest wetland products Petty trade	Domestic use Irrigation Watering livestock Watering trees	70
Poor	Goats 1 Poultry-2 Grass thatched house, mud-walled, flat-roofed house Land < 1 ac Provide casual labor (child labor)	Domestic use	9

season and weather conditions. The diverse crops have different water requirements, and this leads to water conflicts, especially for those produced under irrigation.

Irrigation and General Cropping Pattern

Water withdrawals for irrigation in the Ombeyi watershed are by digging canals from the river, and these lack a proper plan or design. There is also the competition of water for the production of rice and arrowroots (Maranta arundinacea) which have different water requirements. Rice canals are usually shallow since rice is planted in basin-like fields surrounded by bunds, whereas arrowroots are planted in raised areas, and the crop is only flooded occasionally. Arrowroots, unlike rice, are produced throughout the year. Figure 3.6 shows the rainfall for Ahero station which is the closest station to the Ombeyi wetland. The rainfall is highest in April (as shown in Table 3.4), the time when land preparation for rice is carried out in all the three sub-locations. At the start of the cropping season for rice, which in the Ombeyi wetland begins in May with land preparation, a farmer needs to have an irrigation program and a cropping pattern which maximizes economic return and/or water efficiency. In Ombeyi wetland, the existing cropping pattern has been the same for many years and may not utilize water resources

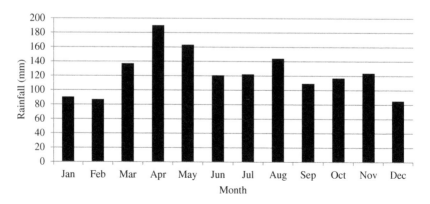

Fig. 3.6 Mean monthly rainfall (mm) for Ahero station from 1950 to 2010

at maximum economic efficiency. Rice is grown only once per year, whereby two or three crops are feasible. Therefore, diversification of the cropping pattern could maximize the net return per unit quantity of land and water. The cropping calendar is not well coordinated, as shown in Table 3.3. All the cropping activities are carried out over an extended period, and this makes water management difficult.

Table 3.3 Rice cropping calendar

Sub-location	Activity	\multicolumn{12}{c}{Month}
		J	F	M	A	M	J	J	A	S	O	N	D
Wang'aya II	Land Preparation					■	■						
	Transplanting								■	■			
	Weeding									■	■		
	Bird Scaring										■	■	
	Harvesting	■											
Kore	Land Preparation					■	■						
	Transplanting								■				
	Weeding									■			
	Bird Scaring	■									■	■	
	Harvesting							■	■				
Ramula	Land Preparation					■	■						
	Transplanting								■	■			
	Weeding									■			
	Bird Scaring									■	■		
	Harvesting		■									■	

Table 3.4 Income and cost of production from selected crops in the Ombeyi wetland

Crop	Income (Kshs)			Costs of production (Kshs)		
	Lowest	Highest	Average	Lowest	Highest	Average
Arrowroots	4800	97,500	35,946	1000	30,000	8892
Maize	3000	90,000	19,341	1100	28,000	5969
Rice	9600	140,000	52,220	1400	40,850	13,707
Sorghum	2500	90,000	21,660	700	23,000	4465
Sugar cane	6000	132,000	36,390	800	43,200	22,122

Source Author

Irrigated Rice

Kenya is a net importer of rice, and between 2006 and 2010 rice from Pakistan accounted for 74% of these rice imports (Short et al. 2012). Rice production is declining due to weak marketing channels and low profitability. Income from the production of rice in Ombeyi wetland computed per 1 ac farm was found to range from Ksh 9600 to 140,000, with a mean of Ksh 52,220. Farm expenditures stood at an average of Ksh 13,800 per 1 ac. Rice farming expenses fluctuated based on the labor cost, agricultural inputs, and source of seeds. The primary costs involved in rice production included land preparation, seeds, transplanting, weeding, scaring away birds, harvesting, and transportation. Since most of the farmers grew one crop of rice per year, and the stated incomes were assumed to reflect their annual income from rice.

Three main rice market chains can be identified within the Kenyan rice sector. These are the integrated large farm chain of Dominion Farms Ltd—an Oklahoma-based corporation operating a large farm in Western Kenya—the traditional market value chain of the non-National Irrigation Board (NIB, a state corporation) irrigated production, and rainfed farmers and the highly concentrated market value chain on the NIB irrigation schemes. Farmers in the smallholder irrigation schemes sell their own rice, and a large number of traders and small-scale rice millers form a local market for the product in the Ombeyi wetland. According to Gitau et al. (2010), the introduction of diesel-powered mills has increased the number of options in the milling industry. In the 1980s, farmers sold most of the rice they produced and relied on maize and beans produced on non-irrigated land for subsistence (Ruigi 1998), but this trend is changing as more farmers now consume their produce.

Sugar Cane

As in the case of rice, Kenya is a net importer of raw sugar as well. In Kenya, each factory has their own cane production land (nucleus estate), but also obtain cane from smallholder farmers/out growers (Monroy et al. 2012). Currently, domestic sugar prices are inflated and are well above the international price for sugar due to tariffs and quotas applied to Kenya's raw sugar importations. While these high prices benefit local producers, they make raw sugar and sugar products more expensive for consumers. It is probable that if the COMESA safeguards were lifted, the inflow of sugar from other countries would drive domestic prices down by about 25%, diminishing the profits of local factories Millennium Cities Initiative (MCI 2008). There are declining returns from sugar cane in the area due to new regulations regarding sugar content, which oblige farmers to invest in improved seeds and crop management. The 2001 Sugar Act requires sugar cane prices that are determined by the sucrose content rather than weight. The reason for this pricing system is to inspire farmers to supply high sucrose sugar cane and millers to improve their sugar recovery ratio, thereby increasing the industry's overall productivity Kenya Sugar Industry (KSI 2009). This condition has discouraged farmers further as they have to acquire new seed varieties. The output of sugar cane per hectare in the 2000s and 1990s has seen a significant decline compared to yields obtained in the 1980s. Possible reasons for this reduction in productivity included the extensive use of low-quality sugar cane varieties, poor agricultural and land management practices, and delayed harvesting of mature sugar cane Kenya Sugar Board (KSB 2010). For sugar cane producers, incomes computed per 1 ac farm range from Ksh 6000 to

132,000 with a mean of Ksh 36,390 with a farm expenditure average of Ksh 22,122.

Arrowroots: A Recent Crop

Arrowroots were introduced into the area in early 1983 by an elderly citizen known as Mr. Awondo, who brought it from Central Kenya where it is planted in the moist river beds. He thought it would be good for the swampy parcels of land that he was not able to utilize for the production of any other crop. He first planted it for subsistence use, but with the growing demand for the crop, it gradually became a commercial crop in 1983 earning him huge sums of money consequently making him very wealthy. Arrowroots thrive where there is enough moisture in the soil for its normal growth and development. The rhizomes can stay for long inside the ground. Arrowroot requires a friable, well-drained, fertile soil; therefore, deep plowing provides favorable conditions for better root development. During the survey period, incomes from the production of arrowroots computed per 1 ac farm were found to range between Ksh 4800 and 97,500, with a mean of Ksh 35, 946. However, results from the KIIs show that with proper crop husbandry the average net income is averagely Kshs 165,400, with a farm expenditure of Ksh 94,600 per 1 ac. Farm expenditures stood at Ksh 94,600 per 1 ac. The high cost of farm expenditures in Ombeyi wetland is due to weeding which is done twice a month during the first five months, depending on the weed population in the field and then once a month for the continuing four months. Irrigation is needed at the early stage of growth especially if the soils do not have adequate moisture. Fertilization increases the yields and the size of the rhizome, but this reduces palatability and the shelf life. Results from KII and FGDs show that these two factors make marketability of rhizomes produced with fertilization difficult. The incomes from arrowroot cited here are based on one season although production is continuous as a new crop is usually planted immediately after harvesting. For this reason, there can be farmers who make approximately Kshs 200,000 per week, when selling the total harvest.

Maize and Sorghum

For the rainfed crops, maize and sorghum, the incomes computed per 1 ac farm ranged between Ksh 3000 and 90,000, with a mean of Ksh 19,341 and Ksh 2500–90,000, with an average of Ksh 21,660, respectively. The farm expenditure for maize and sorghum per 1 ac stood at an average of Ksh 5969 and Ksh 4465, respectively, (see Table 3.4). Figure 3.7 shows the historical trend lines for the production of crops in Ombeyi wetland for arrowroots and rice produced under irrigation and for maize, sorghum, and sugar cane produced under rainfed agriculture.

Use of Wetland for Fish and Papyrus

There are several *natural* wetland products harvested from the Ombeyi wetland. As shown in Fig. 3.8, the most vegetative resource is *Cyperus papyrus*. It is an essential plant to the communities living in the Ombeyi

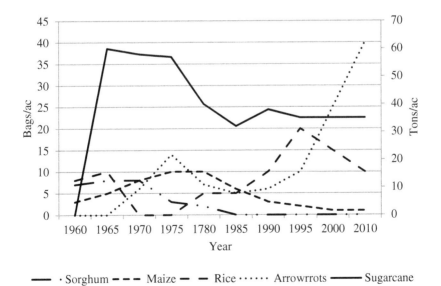

Fig. 3.7 Historical trends for an average production of the 5 crops in the Ombeyi wetland

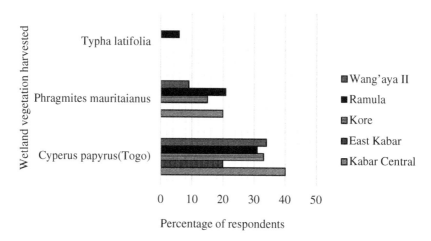

Fig. 3.8 Percentage of respondents harvesting plant wetland resources by sub-location

wetlands and provides an ecosystem service to surrounding communities in the form of building and craft production materials for items such as mats, baskets, chairs, and tables. The plant has several uses depending on the part of the plant and the stage of growth. The most popular harvested stage is when the flower cluster has passed the fan-like stage and is drooping. The papyrus umbel (*oyusi*) is used for making brooms. Papyrus is an excellent source of fuel (Mwanikah 2006), and in some places, the papyrus rhizome (*Osika*) remains the only fuelwood product after the destruction of all other sources (Kibwage et al. 2008). Mature papyrus (*oundho*) is used as a roofing material for houses and is also used as fuelwood.

Fish is another crucial wetland resource for the communities living around the Ombeyi wetlands. The species of fish commonly caught by the respondents include *Barbus nyanzae* (locally called Adel), *Clarius gariepinus* (Mumi), *Clarius werneri* (Nyawino), *Protopterus aethiopicus* (Kamongo), *Oreochromis leucostictus* (Opato), Okong, and *Xenoclaris eupogon* (Ndhira) (See Fig. 3.9). In the past, young fish were not caught and if caught were returned to the water to grow to maturity. However, this is no longer the case. For instance, there is increased harvesting and consumption of fingerlings and juveniles of *Clarius gariepinus* (Nyapus).

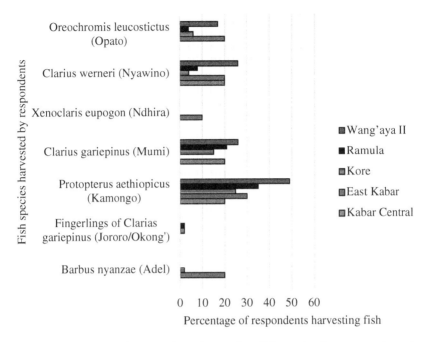

Fig. 3.9 Percentage of respondents harvesting fish wetland resources by sub-location

Factors Affecting Resource Use in the Ombeyi Watershed

We used generalized multiple regression models to analyze the relationship between wetland resource utilization and six predictor variables: number in household, sex of respondent, type of marriage, level of formal education of respondent, age group of the respondent, and type of household. Regression analysis was performed to establish whether the variables 'number in household,' 'sex of respondent,' 'type of marriage,' the 'level of formal education of respondent,' 'age group of respondent,' and 'type of household' are significant factors in resource utilization. The results as shown in Table 3.5 includes information about the quantity of variance that is explained by the predictor variables number in household, sex of respondent, type of marriage, the level of formal education of respondent, age group of the respondent, and type of

Table 3.5 Model summary

Model	R	R square	Adjusted R square	Std. error of the estimate
1	0.354[a]	0.125	0.089	2.99,818

[a]Predictors: (constant), number in household, sex of respondent, type of marriage, level of formal education of respondent, age group of respondent, type of household

Table 3.6 ANOVA[b]

Model	Sum of squares	df	Mean square	F	Sig.
Regression	184.035	6	30.672	3.412	0.004[a]
Residual	1285.439	143	8.989		
Total	1469.473	149			

[a]Dependent variable: Resource utilization
[b]Predictors: (constant), number in household, sex of respondent, type of marriage, level of formal education of respondent, age group of respondent, type of household

household. In this model, the R value is 0.354[a], which indicates that there are lots of variances shared by the predictor variables and resource utilization. The R square value is 0.125, which shows that 12.5% of the variance in the dependent variable is explained by the independent variables in the model.

The F statistic as shown in Table 3.6 represents a test of the hypothesis whether the R square proportion of variance in the dependent variable accounted for by the predictors is zero. It appears that the six predictor variables in the present study are not all equal to each other and could be used to predict the dependent variable, resource utilization, as is indicated by a significant F value (3.412) and a small significance level (0.001). This indicates that for resource utilization, the level of formal education of respondent, sex of respondent, type of marriage, age group of respondent, and type of household have a role to play, as discussed below.

The level of formal education of respondents is essential, as those with formal education can get employment and thus reduce the dependence on natural resource dependence. The higher the percentage of illiterate people, the higher the dependency on livelihoods based on natural resources and consequently the inability for trade-offs, which potentially reduces the opportunities available for them regarding livelihoods strategies.

Sex of respondent—There are apparent biases regarding resource utilization between female-headed and male-headed households, with differential ownership of, access to, and use of natural resources for men and women, mainly due to institutional restrictions for women.

Type of marriage—The type of marriage, i.e., monogamous or polygamous marriage, affects resource use, as in a polygamous marriage resources may be split among the households, competing for needs, ultimately narrowing scenarios for trade-offs.

Age group of the respondent—On the one hand, the number of child-headed homes is increasing, with orphans living in desperate conditions, not only because they were born into needy families but also because they are still too young and lack financial and entrepreneurial ability to manage properties left to them by their departed parents. The vulnerability may limit their ability to make decisions on the trade-offs to make, as this would jeopardize their livelihoods. On the other hand, the elderly are the relatively conservative voices of society and their knowledge of traditional systems may hinder their understanding of the changes taking place, although it may help in decision-making based on what has worked well previously.

Type of household—The impact of the HIV/AIDS pandemic has left many women in their mid-20s to 40s as household heads, making them vulnerable due to the shock of being financially, psychologically, emotionally, and physically unprepared. They are disadvantaged, as their ownership or access to land is not guaranteed, and hence, investing in land-use changes as a result of trade-offs may be difficult.

Table 3.7 provides more information about the effects of the variables number in household, sex of respondent, type of marriage, the level of formal education of respondent, age group of respondent, and type of household on resource utilization. As shown in the table, the unstandardized coefficients for the number in household, the level of formal education of respondent, and age group of the respondent in this case are 0.287, 0.014, and 0.321, which indicates that for the predictor variables, resource utilization will be affected by, 0.287, 0.014, and 0.321,

Table 3.7 Coefficients[a]

Model	Unstandardized coefficients		Standardized coefficients	t	Sig.
	B	Std. error	β		
(Constant)	10.142	1.541		6.581	0
Type of household	−0.417	0.564	−0.069	−0.738	0.462
Age group of respondent	0.287	0.168	0.143	1.705	0.09
Sex of respondent	−1.348	0.591	−0.213	−2.28	0.024
Type of marriage	−0.134	0.399	−0.026	−0.335	0.738
Level of formal education of respondent	0.014	0.022	0.051	0.644	0.521
Number in household	0.321	0.137	0.189	2.345	0.02

[a]Dependent variable: Resource utilization

respectively. Examining the beta coefficients for the number in household, the level of formal education of respondent, and age group of the respondent, it can be noted that these three variables are more naturally the better predictors of resource utilization.

Examining the t statistic for the variables, it can be seen that they are associated significance values of 0.090, 0.521, and 0.020, indicating that the null hypothesis, which states that this variable's regression coefficient is zero when all other predictor coefficients are fixed to zero, can be rejected. This shows that the resource utilization can be predicted by variables number in household, the level of formal education of respondent, and age group of the respondent.

In other words, it appears that there is a statistically significant relationship between independent and dependent variables. Therefore, it can be concluded that resource utilization is dependent on the number of persons in the household, the level of formal education of respondent, and age group of the respondent.

The unstandardized coefficients indicate the increase in the value of the dependent variable for each unit increase in the predictor variable. In this case, the unstandardized coefficients for sex of respondent, type of marriage, and type of household are −1.348, −0.134, and −0.417, which indicates that for each of these variables, predicted resource utilization would decrease by −1.348, −0.134, and −0.417. Examining the beta coefficient for sex of respondent, type of marriage, and type of household, we can see that these variables are not better predictors of resource utilization. Examining the t statistic for these variables (−2.280, −0.335, and −0.738), it can be seen that they are associated with high significance values of (0.024, 0.738, and 0.462), indicating that the null hypothesis, which states that this variable's regression coefficient is zero when all other predictor coefficients are fixed to zero, can be accepted. This shows that sex of respondent, type of marriage, and type of household are not better predictors of resource utilization.

Thus, it can be concluded that number in the household, the level of formal education of respondent, and the age group of respondent are better predictors of resource utilization. The sex of respondent, the type of marriage and the type of household are not good predictors of resource utilization as they have a negative relationship but since they are categorical or nominal variables they cannot have an inverse relationship in which the increase of one variable causes a decrease in the other variable.

The nature of household use of wetlands strongly differentiated across households and appeared to be profoundly influenced by socioeconomic factors. Female-headed families with a big size were more likely to engage in both wetland cropping and collection of natural wetland products. Access to income from off-farm activities, land holding size per capita, and wealth significantly relate to use of wetland for cropping and collection of natural products. Households with access to irrigation plots above 2 acres were less likely to engage in the collection of natural products as expected but more liable to participate in wetland cropping (Table 3.8).

The Impact of Recent Changes

Technological Changes

Highland arrowroot technology is a modern agriculture technology in Kenya which allows the growing of the crop where there is no river or wetland. The use of highland arrowroot technology is a significant development, in modern agriculture technology in Kenya introduced

Table 3.8 Number of respondents doing irrigation and collecting wetland products

		Do your household members harvest any natural products from the wetlands?		Total
		Yes	No	
Do you do irrigation?	Yes	83	53	136
	No	5	9	14
	Total	88	62	150

by KALRO Katumani—which is one of the national research stations mandated with agricultural research, including crop development. The method is switching from the tradition of growing the crop only in the river valleys to growing the crop in upland areas. One acre accommodates 29,333 plants which yield one tuber each giving a total of 29,333 tubers sold at Ksh. Ten gives Ksh 293,333 gross income per acre in 6–8 months. Today indigenous tubers are in the class of high-value crops in Kenya and enjoy good market prices as a consequence of increased intake of the tubers for breakfast and as snacks, due to higher consumer awareness of healthy eating habits. This has led to rapid growth in the urban market share for the root crops.

Arrowroots are grown in the river valleys where there is plenty of water, but this is changing as many river valleys have since dried up, seemingly due to global warming and subsequently climate change. The use of modern agriculture technology in Kenya, especially highland arrowroot technology, is on the rise as a climate adaptation method. The work entails planting the germplasm in trenches 60 cm deep, 1 m wide, and any desired length. The trench is lined with thick gauge polythene paper and filled with the soil–manure mixture at a ratio of 2:1 or one wheelbarrows of manure per meter square. As a recent agricultural technology with new prominence, it involves excavating trenches spaced at 0.5 meters. The suckers are planted at a spacing of 30 cm by 30 cm by burying the germplasm up to 20 cm, followed by thorough watering. When planting is done during the dry season, the rainy mulching is recommended. Watering should be done once a week to maintain wetness with the old and dry leaves regularly removed and to keep the plot free of weeds. The crop is ready for harvesting in a period of 6–8 months (Agro-Environment Initiative 2014).

Discussion

The ecosystem approach demands robust stakeholder participation involving all those who have an interest in or could be affected by, decision-making. Involvement in decision-making is crucial because the ecosystem approach is about managing human activities. People are much more likely to act upon a resolution and change their behavior if they approve the basis on which it was made, and they comprehend it only when there is full and active participation. Stakeholders were engaged in the scenario study to provide opinions on the different scenario features on land use. The land-use distributions for different scenarios were based on the percentage of rice, sugar cane, arrowroots, and papyrus per unit area. Discussion with the community on the crops grown and the prices obtained enables us to come up with the trade-offs given different scenarios.

The farmers and wetland resource users can take steps to ensure their activities conform to the policy if they understand the problems and possible solutions and encourage others to do so. The civil society has a significant role to play, through lobbying, advocacy, research, and innovation on issues that promote the wise use of wetlands. The nature of most of the civil society organizations within the area is humanitarian. Resource users can contribute to wetland conservation by making more sustainable decisions regarding their consumption and resource use and by positively influencing others through advocacy, petitioning, reporting of non-compliance, and legal action.

From the stakeholder analysis, three main trade-offs were identified: *the first one* was between crop production and natural production of fibers for livestock grazing, mat making, and building. The papyrus is produced for making mats, and Phragmites is used for fencing and roofing, whereas the livestock feed on different types of grasses found in the wetlands. Livestock is also grazed in the rice fields after the crop has been harvested. Results show that the expansion of the cropping fields is not away from the wetlands but into the wetlands. Papyrus is usually uprooted or burnt to clear the field for crop production. This reduces the percentage of wetland vegetation and also affects the livelihoods of those dependent on them. The most significant trade-off between crop production and natural production of fibers in arrowroot production as this targets the virgin soils of freshly cleared lands which are frail and high in nutrients.

The *second* trade-off was between crop cultivation and hydrological regulation. The local population has little awareness about the role of wetlands in hydrological regulation, whereas this is a concern of environmentalists and department staff in organizations such as NEMA and WRMA that give it a high priority. The local community reported increased runoff during rains often causing crop destruction, soil erosion, flooding, and destruction of property. Shallow wells are also reported to dry up during the dry season, something that was not witnessed in the past. Wetland vegetation traps sediments and slows down the speed of water. Cultivation has reduced the water recharge function of the Ombeyi wetlands.

The *third and last* trade-off was related to the reduction in soil organic matter associated with the artificial drainage of wetland farms and unsustainable agricultural practices. This practice seemed in conflict with the future use of the wetland for crop production. The introduction of arrowroot production on a large scale and the type of agronomical practices carried out for the crop has led to depletion of soils. Arrowroots are produced all year round with no time given for the soil to replenish nutrients as the replanting is done the same time as harvesting and this goes on until the production reduces to uneconomical levels. The drainage for arrowroots plots is more profound than that of rice plots which means there is a faster flow of water. Weeding of arrowroots is done at least 14 times in nine months. The weeding regime is twice every month for the first five months and then once a month for the remaining four months which increases the costs of labor, reduces the profit margin, and also removes organic matter.

Farming has increased the input of water pollutants and removed the natural water filtering services provided by wetlands. This situation is also threatening livelihoods based on the natural production of wetland fibers on which the poor in particular rely and may, in fact, increase poverty for some groups, especially papyrus harvesters.

Recommendations

On the basis of the above analysis, some recommendations could be made to help suppress the coming crisis in wetland management and to enhance sustainability practices. Among them are: improving wetland management through awareness creation, capacity building, and programs aimed at supporting alternative livelihood options of the poor to reduce pressure on wetland resources. In the Ombeyi wetland, introduction of improved agricultural technologies, investment in irrigation infrastructure, improving access to markets, and promotion of alternative livelihood strategies should be promoted. Opportunities identified together with communities for broadening the livelihood base include ecotourism, identifying new markets for off-farm income, increased access to high yielding varieties, improving access to markets, and better extension services. Other recommendations are the zoning of arrowroot, paddy rice, and sugar cane cultivation areas. Arrowroots and paddy rice should be restricted to areas of seasonal wetland, to eliminate the need for draining the wetland for cultivation. Sugar cane should only be grown in the temporary wetland areas. The swamp should only be used for the sustainable harvesting of bioresources.

Note

1. As defined by the Canadian Ministry of Natural Resources (http://www.web2.mnr.gov.on.ca/mnr/biodiversity/wetlands/what_are_wetlands.pdf, accessed October 20, 2017). The four common types of wetland are: marshes, swamps, bogs, and fens (ibid.).

Appendix

See Appendix Table 3.9

Table 3.9 Schedule of meetings held with stakeholders

Date	Site	Activity	No. of participants
15 November 2013		Inception workshop	29
22 January 2014	Ramula sub-location	Community awareness creation meeting—Kiliti church	
24 January 2014	Kore sub-location	Community awareness creation meeting.	32
6 February 2014	Kore sub-location	Community awareness meeting	27
12 February 2014	North East Kano (Oroba) sub-location	Awareness creation meeting at Oroba church	25
19 February 2014	Wangaya II sub-location	Community awareness meeting at Sanda primary school	22
8 April, 10 April, 13, 14, and 19 May 2014	Wangaya II sub-location	Focus group discussion/community mapping	14
15, 16, 21–23 May 2014	Kore sub-location	Focus group discussion/community mapping	14
26–30 May 2014	Ramula- Kiliti church	Focus group discussion/community mapping	14
4–6 June 2014	Kore, Ramula, and Wang'aya II sub-locations	Key Informants Interviews	6
11 June 2014	Kore, Ramula, and Wang'aya II sub-locations	Scheme leaders meeting	18
13–26 June 2014	NEMA, LVEMPII, KWS, KFS, Dept. of Agriculture, Dept. of Fisheries, Dept. of Planning, NIB County Executive Member for Water, KALRO Chiefs, Assistant Chiefs	Key Informants Interview	14
4 December 2014	Ombeyi/North East Kano	Community feedback workshop—Wi-pinje Eco-resort center	44
5 December 2014	Kisumu hotel	End of project stakeholders workshop	32

References

Agro-Environment Initiative. (2014). *Modern agriculture technology in Kenya: Highland arrowroot technology can earn Ksh 293, 330*. Accessed June 2, 2016, from http://yagrein.blogspot.com/2014/03/modern-agriculture-technology-in-kenya.html.

Brechin, S., Wilshusen, P., Fortwangler, C. L., & West, C. (2002). Beyond the square wheel: Toward a more comprehensive understanding of biodiversity conservation as a social and political process. *Society and Natural Resources, 15*, 41–64.

Cohen-Shacham, E., Dayan, T., Feitelson, E., & De Groot, R. S. (2011). Ecosystem service trade-offs in wetland management: Drainage and rehabilitation of the Hula, Israel. *Hydrological Sciences Journal, 56*(8), 1582–1601.

Darradi, Y., Morardet, S., & Grelot, F. (2006). *Analysing stakeholders for sustainable wetland management in the Limpopo River Basin*. 7th WaterNet/WARFSA/GWP-SA Symposium, 1–3 November. Lilongwe, Malawi: IWMI.

Davis, J., & Froend, R. (1999). Loss and degradation of wetlands in south-western Australia: Underlying causes, consequences and solutions. *Wetlands Ecology and Management, 7*(1), 13–23.

Galbraith, H., Amerasinghe, P., & Huber-Lee, A. (2005). *The effects of agricultural irrigation on wetland ecosystems in developing countries: A literature review* (CA Discussion Paper 1). Colombo: Comprehensive Assessment Secretariat.

Gitau, R., Mburu, S., Mathenge, M. K., & Smale, M. (2010). *Trade and agricultural competitiveness for growth, food security and poverty reduction: A case of wheat and rice production in Kenya* (Working Paper Series 45/2011). Nairobi: Tegemeo Institute of Agriculture and Policy and Development.

Gren, I.-M., Folke, C., Turner, R. K., & Bateman, I. (1994). Primary and secondary values of wetland ecosystems. *Environmental & Resource Economics, 4*, 55–74.

Kibwage, J. K., Onyango, P. O., & Bakamwesiga, H. (2008). Local institutions for sustaining wetland resources and community livelihoods in the Lake Victoria basin. *African Journal of Environmental Science and Technology, 2*(5), 97–106.

KSB. (2010). The Kenya sugar industry value chain analysis: Analysis of the production and marketing costs for sugarcane and sugar related products. In L. Monroy, W. Mulinge, & M. Witwer (Eds.), *Analysis of incentives and disincentives for sugar in Kenya*. Rome: MAFAP, FAO, Technical notes series.

KSI. (2009). *Kenya sugar industry strategic plan 2010–2014*. Retrieved May 2, 2017, from www.kenyasugar.co.ke/downloads/KSI%20Strategic%20plan.pdf.
McCartney, M., Rebelo, L.-M., Senaratna Sellamuttu, S., & De Silva, S. (2010). *Wetlands, agriculture and poverty reduction* (IWMI Research Report No. 137, 39 pp.). Colombo, Sri Lanka: International Water Management Institute. https://doi.org/10.5337/2010.230.
MCI. (2008). *Sugar in Kisumu, Kenya*. New York: The Earth Institute at Columbia University.
MEA. (2005a). *Ecosystems and human well-being: Synthesis*. Washington, DC: Island Press.
MEA. (2005b). *Ecosystems and human well-being: Wetlands and water synthesis*. Washington, DC: World Resources Institute.
Monroy, L., Mulinge, W., & Witwer, M. (2012). *Analysis of incentives and disincentives for sugar in Kenya*. Rome: MAFAP, FAO, Technical notes series.
Mwanikah, M. O. (2006). *Sustainable use of papyrus cyperus papyrus at Lake Victoria wetlands in Kenya: A case study of Dunga and Kusa swamps, papyrus swamps along Lake Victoria*. Monterrey: Sustainability Institute, Tecnologico de Monterrey, Mexico.
Reed, M. S. (2008). Stakeholder participation for environmental management: A literature review. *Biological Conservation, 141*(10), 2417–2431.
Ruigi, G. M. (1998). *Large-scale irrigation development in Kenya past performance future prospects* (Food and Nutrition Planning Unit Report No. 23). Nairobi: Ministry of Planning and National Development.
Schmidt-Soltau, K. (2004). The costs of rainforest conservation: Local responses towards integrated conservation and development projects in Cameroon. *Journal of Contemporary African Studies, 22*(1), 93–117.
Senaratna Sellamuttu, S., De Silva, S., Nguyen-Khoa, S., & Samarakoon, J. (2008). *Good practices and lessons learned in integrating ecosystem conservation and poverty reduction objectives in wetlands*. Colombo: International Water Management Institute—Wageningen: Wetlands International, 73 pp., on behalf of, and funded by, Wetlands International's Wetlands and Poverty Reduction Project.
Short, C., Mulinge, W., & Witwer, M. (2012). *Analysis of incentives and disincentives for rice in Kenya*. Rome: MAFAP, FAO, Technical notes series.
Sunderland, T. C. H., Ehringhaus, C., & Campbell, B. M. (2008). Conservation and development in tropical forest landscapes: A time to face the trade-offs? *Environmental Conservation, 34*(4), 276–279.

van Dam, A., Kipkemboi, J., Rahman, M., & Gettel, G. (2013). Linking hydrology, ecosystem function, and livelihood outcomes in African papyrus wetlands using a Bayesian network model. *Wetlands, 33*(3), 381–397.

Wanzie, C. S. (2003). Wetland conservation and development in the Sahel of Cameroon. Jean-Yves Jamin, Lamine Seiny Boukar, Christian Floret. 2003, Cirad—Prasac, 6 pp. Online at: https://hal.archives-ouvertes.fr/hal-00137946. Accessed December 5, 2016.

Wood, A., & van Halsema, G. E. (2008). *Scoping agriculture-wetland interactions: Towards a sustainable multiple-response strategy* (FAO Water Report No. 33). Rome: FAO.

4

Agriculture, Ecology and Economic Development in Sub-Saharan Africa: Trajectories of Labour-Saving Technologies in Rural Benin

Georges Djohy, Honorat Edja and Ann Waters-Bayer

Introduction

Improving economic growth in Benin Republic is based primarily on developing agriculture, which employs about 70% of the active population, contributes up to 36% of GDP (Gross Domestic Product) and provides about 88% of export earnings (PSRSA 2011). The agricultural policy of the country over recent decades was based on two main pillars: (i) GDP growth and sustainable development and (ii) food security and subregional solidarity.

G. Djohy (✉) · H. Edja
Faculty of Agricultural Sciences, University of Parakou, Parakou, Benin

A. Waters-Bayer
Royal Tropical Institute (KIT), Amsterdam, The Netherlands
e-mail: waters-bayer@web.de

© The Author(s) 2018
J. Abbink (ed.), *The Environmental Crunch in Africa*,
https://doi.org/10.1007/978-3-319-77131-1_4

Regarding the first pillar, the cotton sector has benefitted first and foremost from the attention of the state, as it is the country's prime export crop and main source of foreign currency. More recently, ideas and some timid efforts to promote other promising cash crops, such as cashew nut, pineapple and oil palm, have been made so as to diversify the sources of foreign exchange beyond the cotton sector, which is experiencing many difficulties in recent times.

Regarding the second pillar, the State has been promoting crop diversification with particular emphasis on rice, maize, yam and cassava, all of which are staple foods for the Benin population. Beyond national food self-sufficiency and sovereignty, these food crops also feed various forms of national symbolism and demonstrate power within the framework of practices and discourses linked to solidarity within the subregion of West Africa. In recent years, Benin has repeatedly demonstrated some degree of 'charity' towards neighbouring countries in various situations of food deficit or crises due to severe droughts and other shocks reportedly related to climate change. The generosity of the Benin Government and the policy towards the people of Niger have fuelled good relationships between the two countries. Individuals' and media testimonies of food donation ceremonies between Benin and Niger have reported, for example:

> Benin has decided to donate one hundred and fifty (150) tons of maize to its neighbour Niger, in solidarity with that country with which it shares a border, and which is currently plagued by a food crisis that threatens millions of people. (Ouestaf News 2010)

> Benin Government has donated 150 tons of local white maize to Niger, whose population has been suffering from a severe food crisis caused by climate hazards. […]. Benin, in a surge of solidarity with this neighbouring country with which it maintains privileged relations, has a duty to give assistance to its hard-hit people. (Maina 2010)

> Benin Government has decided to support the people of Niger through a donation of 150 tons of maize to help this brother and friend country passing through this difficult situation. During the donation ceremony, the Head of the Benin delegation expressed to the Nigerien authorities the

satisfaction of Benin authorities in fulfilling a duty of solidarity towards a neighbouring country linked to Benin by longstanding friendly and fraternal relations. (Kinninvo 2010)

The shift from using maize for local food to using maize to create and maintain subregional solidarity links between countries reflects the changes in the politico-institutional environment in which new agricultural transformation and development policies have emerged in Benin over the last few decades.

Agricultural mechanisation and modernisation, although sometimes elusive concepts used for political and electoral purposes, bring about real change in agricultural and agrarian systems at the grass-roots level. Already in the 1990s, Benin initiated a change in its agricultural production system, introducing tractors within the framework of North–South technology transfer supported by various European partners, including the French Development Agency (AFD) and the French Farmer and International Development (AFDI). The advent of CUMAs (Farm Equipment User Cooperatives) is one of the most important technological successes in the agricultural sector since the introduction of draught animals in the 1950s, together with the promotion of cash crops, in particular cotton. The precursors of the CUMAs were the cooperative groups, with rather disappointing results, set up in 1961 by the Union of Swiss Cooperatives and meant to ensure ploughing, spraying and sowing through tractors (Addrah 1979). A CUMA is a local cooperative composed of about ten smallholder farmers equipped with a 30–70 hp tractor, a plough and a 3-ton trailer, requiring a total budget of 10 million XOF (US$20,000) (Balse et al. 2015; Djohy 2017).

This technical-institutional model of access to and use of agricultural equipment, which has been successful in various municipalities in the country, was adopted and retained as reproducible and generalisable through the Programme for the Promotion of Agricultural Mechanization (PPMA, later renamed the Agricultural Mechanization Development Agency or ADMA). The PPMA at the time of fieldwork was the entity through which the government implemented its overall policy of modernisation of the agricultural sector, as reflected in the Strategic Plan for Agricultural Sector Development (PSRSA) and the

National Agricultural Mechanization Strategy Document (DSNMA) and embodied at the institutional level in the National Council of Agricultural Mechanization (CNMA) set up in 2007 (JORB 2008). The PPMA approach consisted of providing agricultural equipment, especially tractors, subsidised at 50% of the real price, to eligible individuals, farmer groups and cooperatives and young 'agripreneurs', to be repaid in four instalments spread over four years. For example, 30 hp and 60 hp tractors were sold to farmers at 4.5 million (US$9000) and 6 million XOF (US$12,000), respectively, 20% of which were paid at the time of equipment collection, 30% each in the second and third years and 20% in the fourth year (Saizonou 2009). Through the PPMA, several hundred farmers have thus been equipped with tractors and other accessories. The equipment-supply mechanism and all associated facilities—both local and external tractor service delivery having developed in rural areas—have greatly improved the level of access of smallholder farmers to motorised services, especially for ploughing.

Although the available statistics may be subject to the dynamics of 'data arrangement' that have prevailed in public arenas in Benin in recent years (see Fichtner 2016) or the engineering of figures as a strategy of showcasing success and lobbying decision-making (Djohy 2017), it is obvious that modernisation is gradually taking place, with small-scale farmers increasingly using equipment imported from China, India and elsewhere. Despite the reality of recurring breakdowns of machinery, limited availability of spare parts and lack of local expertise for repair and maintenance of equipment, which are major constraints to the development of mechanisation (Saizonou 2009), it must be acknowledged that the tractor is becoming increasingly integrated in local cropping systems in Benin.

In addition to tractors, *pesticides* in general and *herbicides* in particular are the second category of technologies associated with the major changes in the agricultural sector. The history of pesticides dates back to the introduction of cotton cultivation in Benin, like in many other African countries, as a result of the antecedent conditions of blocking French cotton in American ports during the Civil War of 1861–1865, which prompted French industrialists from 1895 onwards to consider diversifying their sources of raw material by turning towards French West

Africa in order to guarantee and secure the needs of the French textile industry (Kpadé and Boinon 2011). Also labour-saving technologies for controlling various pests were successfully tried and used in the form of insecticides and fungicides. Herbicides were introduced in Benin in the 1990s, after experiments on crops such as groundnut, cotton, maize and sorghum in a subregional programme involving other francophone African countries such as Cameroon, Mali and Côte d'Ivoire (Atachi 1979; Gaborel and Fadoegnon 1991; Marnotte 1994, 1995).

Although these chemicals were adopted and used since their introduction into Benin's farming landscape, the government's input-supply policy does not allow easy access by all farmers. In practice, formal flows of highly subsidised inputs are made available only to cotton farmers, who receive the inputs on credit at the beginning of the agricultural season; the input costs are deducted from their overall income after they sell the harvest. All other farmers, particularly those who do not grow cotton, access crop inputs only informally and through their social networks. Inputs are sometimes sold by cotton farmers to others as part of their community-based solidarity and used through various practices non-compliant with the recommendations of the extension services (Kherallah et al. 2001). This makes the dichotomous classification of farmers with regard to their access to inputs something totally fuzzy because of the realities on the ground.

The gap between users and non-users of chemical inputs completely disappeared after 2010, when the Government reformed the cotton sector, which was faced with various difficulties and conflict of interests between the actors involved. The lack of agreement between parties and the politico-institutional crises that arose from the State's desire to reorganise the cotton sector, supposedly to increase its contribution to the national economy, led in May 2012 to the de-liberalisation of the sector by the State, which unilaterally cancelled all the liberalisation measures taken in the 1990s (CTA 2008; Meenink 2013a). This return to the 'all in the hands of the authorities'-model and the abuses arising therefrom paved the way for the emergence of enormous informal flows of chemicals dumped onto the local market by private operators (Meenink 2013b). Although chemicals have always circulated informally in the local market, their almost official reign today is the outcome of the breakdown in public–private partnership and thereby in the process of liberalisation of the agricultural sector started from the 1992–1993 agricultural season.

In this changing political and institutional context, in which various ambitions, policies and reforms intertwine in connection with promoting economic growth and development, our study aims at understanding the trajectories of the various labour-saving technologies introduced or promoted at the local level. From tractors to pesticides, Benin's agriculture has undergone major transformations, although very few studies focus on and trace them from an anthropological perspective. Apart from some recent work on CUMAs (Gibigaye 2008; Gibigaye et al. 2010; Moumouni et al. 2013; Balse et al. 2015) and those carried out on pesticides (Agbohessi et al. 2012; Adechian et al. 2015; Ligan-Topanou et al. 2015), no other published study addresses the broad impacts of the mechanisation and pesticidation of Benin agriculture. In this chapter, we hypothesise that, even if the State has the key to the promotion of the above-mentioned agricultural technologies with its logic of promoting economic growth and agricultural development, the possible pathways of these technologies at grass-roots level remain unpredictable and out of control. As a result, new agricultural technologies not only reinforce the polarisation of rural livelihoods, but also generate new land-use practices that are quite environmentally unfriendly and bear socio-economic impacts that could be very costly in the long run.

Theoretical Framework

Our scrutiny of the trajectories of agricultural labour-saving technologies in northern Benin has been fundamentally inspired by political ecology as well as science and technology studies that have occupied an important place in socio-anthropological analyses over recent decades.

Political ecology is not a theory as such, but rather a subfield which 'emphasizes how power relations and politics shape the dynamics of economic development, environmental transformation, and social change across geographic scales of analysis from the local to the global' (Jarosz 2001, 5472). This perspective is rooted in the premise that ecological changes for a large part are products of political processes, and it seeks to understand the complexity of the relationships between communities and their environment through the various forms of access to and control over natural resources within the framework of their livelihoods' sustainability (Watts 2000; Robbins 2012). In this study, the use of a political ecology perspective has allowed deeper insight into the interplay between socio-economic, political and ecological processes linked to the transformation of livelihoods and the evolution of interactions between rural actors. In addition, emphasis was put on technology as inspired by science and technology studies, since technological innovations in either hard or soft versions are not often simply adopted, but rather appropriated and repurposed to accommodate specific interests of relevant actors and their interactions with others. Technology itself has been understood in its broadest sense as all means by which human beings control or transform their natural environment (Spier 1970).

We do not go into details of the social processes related to the design, manufacture and development of these technologies, but rather consider the technologies in use in order to understand how they and societies in which they were introduced and promoted are reciprocally shaped. We embrace the view that technology and society cannot be considered as separate entities simply influencing each other, but that the rural world as we have seen and studied it is a product of various social, economic, financial and political factors reflecting a socio-technological configuration at a particular moment in time (Hughes 1983; Mackay and Gillespie 1992; Law 2012). This is an assemblage of heterogeneous elements forming a network representing a new socio-technological order that comes to topple the existing order and promotes a reconfiguration of power relations (Vinck 2012).

Combining political ecology with a reinterpretation of agricultural technologies in the hands of their users amounts to considering that a given technology promoted at a particular time in a given environment may receive new meanings and uses intended to change the course of existing power relations between local actors. In the case of herbicides, for example, manufacturers, importers/distributors, governments and agricultural development organisations promoting them unanimously agreed on their main function of systemic or selective elimination of weeds that compete with crops and negatively affect agricultural yields. Hand-weeding has always been perceived as one of the bottlenecks in the agricultural calendar (Marnotte 1994, 1995). Other forms of use of these products on the ground could fall under the interpretive dynamics of users, who are also important actors in innovation (Akrich 1998). The increased use of herbicides could be the result of what farmers are able to achieve with these products beyond the weed control for which the products are supplied. They could use the products to increase their power over land, negotiate more space or take advantage of their coexistence with other socio-professional groups in the community. The development of informal pesticide-supply chains could also be stimulated not only by a fragile institutional environment as described above, but also by a saturation of the labour market, obliging young unemployed graduates to seek new survival opportunities. From the manufacturers to the end users, the trajectories of agricultural chemicals are numerous and diverse and probably well integrated into existing power relations. Latour (1986, 267), corroborating the unpredictable paths of technological innovations, ended by asserting that 'The spread in time and space of anything—claims, orders, artifacts, goods—is in the hands of people; each of these people may act in many different ways, letting the token drop, or modifying it, or deflecting it, or betraying it, or adding to it, or appropriating it'.

Research Setting

Physical Environment

The study was carried out in the district of Gogounou, situated at the southern entrance of Alibori Province, the largest pastoral region of Benin. It is located between 10°33' and 10°57' North latitude and 2°15' and 3°15' East longitude, with an area of 4910 km². Gogounou District experiences a Sudano-Guinean climate characterised by a rainy season from May to October and a dry season from November to April. The annual average rainfall is 1100 mm, with most of it between August and September. The temperature varies during the year between 18 and

38 °C; the cold and dry harmattan is experienced from November to February (PDC$_1$ Gogounou 2004).

The district lies within the Niger river basin and has two main rivers, Sota in the east and Alibori in the west, with various tributaries. It consists essentially of plains and plateaus with some hills rising up to 300 metres. The soils are of granite-gneissic base, mostly ferruginous and generally suitable for agriculture. In the alluvial plains, the clayey-sandy soils are fairly rich in organics brought in by annual flooding of the rivers (PDC$_2$ Gogounou 2010).

The cultivable area in Gogounou District is estimated at 1705 km^2, about 35% of the total area. Two classified forests, namely *Trois-Rivières* in the east and *Alibori Supérieur* in the west, cover about 177,200 ha or 36% of the total area of the district. Pastures and lowlands cover 123,500 ha (25%) and 360 ha (0.07%), respectively, of the district's area. Along the watercourses are forests with large trees such as *Khaya senegalensis, Pterocarpus urinaceus, Afzelia africana, Adansonia digitata* and *Ceiba pentandra* (PDC$_2$ Gogounou 2010).

Human and Social Context

The human population of Gogounou is composed of three main groups: the Baatonou also called Bariba (49%), the Fulani (44%) and the foreign nationals and internationals (7%). Islam is the dominant religion practised by 82% of the local population, followed by Christianity (12%). Traditional religions and other types of beliefs and ideologies are practised by 6% of the population (INSAE 2016; PDC$_3$ Gogounou 2017).

Gogounou experiences a fairly *rapid population growth*: from 50,045 inhabitants in 1992 to 80,013 inhabitants in 2002 (PDC$_1$ Gogounou 2004). The most recent census in 2013 revealed a population of 117,523 inhabitants (58,018 men and 59,505 women), divided into 15,250 households of eight people on average (INSAE 2016). History reveals that the Bariba people were the first to cultivate the area and are landowners who receive land as common heritage managed by traditional chieftaincy. Land is not meant to be sold except in peri-urban areas, where recent subdivision and parcelling have resulted in sales of plots (PDC$_1$ Gogounou 2004).

Agriculture is the most important economic activity in Gogounou and is the prerogative of the Bariba, who cultivate cash crops including cotton and food crops including maize, millet, sorghum and rice. Agricultural holdings are continuously increasing and farm workers have been recently estimated at 39,994 individuals (19,945 females and 20,049 males). A total of 2433 small-scale farms, of which 415 (17%) were headed by women and 2018 by men, were identified in Gogounou during the 2015–2016 agricultural campaign (PDC$_3$ Gogounou 2017).

Gogounou is also an important livestock-keeping area, with cattle, small ruminants and poultry being the main species kept. Livestock is the primary economic source for the Fulani pastoralist households. A total of 77,970 head of cattle, 29,057 sheep, 22,826 goats and 80,658 poultry was found in Gogounou in 1999, which corresponded to a value of over five billion XOF (US$10 million) (PDC$_1$ Gogounou 2004). However, in recent years, livestock keeping especially by Fulani pastoralists has faced a much less favourable situation. Ecologically unfavourable conditions and agricultural expansion, to be detailed in this chapter, have contributed to a reduction in the local ruminant population.

Research Methods

We first sought to understand from a GIS-based diachronic analysis the land-cover change in Gogounou District during the last three decades. When we had a clear view of the evolutionary dynamics of various spatial units from satellite images provided by the National Remote Sensing Center (CENATEL), we then sought 'to link the people to the pixels'. In reality, merging remote sensing with social science analysis starts from the assumption that remotely sensed images are not mere representations of nature in a given ecological context, but carry covert complex sociopolitical realities (Liverman et al. 1998; McCusker and Weiner 2003).

To uncover this hidden political ecology, we used participant observation as the main technique for collecting ethnographic data. In Bariba farms, we followed and participated in all cropping operations. With

Fulani pastoralists, we used the same technique in trying to understand the uses they make of new agricultural technologies, their perceptions of the uses by other actors and the potential impacts on their livelihoods. These two groups of actors and other resource persons were engaged in open-ended and semi-structured interviews, as well as group discussions to better understand social processes related to labour-saving technologies, especially herbicides, widely used throughout Gogounou. The study mobilised 164 individual interviews and 21 group discussions during ten months of fieldwork in the period from July 2013 to October 2014. Interviews were recorded, transcribed and analysed using F4 software. For the purpose of this chapter, some other updated data were mobilised from various sources (mostly extension service and online databases) to support the analysis.

Results

More Tractors, More Pressure on Land

Gogounou District is one of the regions that have benefited from the policy of agricultural mechanisation initiated by the Benin Government. To date, there are nearly 100 tractors used in various agricultural operations, mainly ploughing. The equipment consists of cultivators and tractors of different capacities (Table 4.1 and Photo 4.1). Fifteen CUMAs have been set up in Gogounou in recent years and nine of them were equipped by the PPMA. Although these cooperatives own and operate their tractors, agricultural equipment in the municipality remains mainly the property of individual farmers who also acquired them through the PPMA (Ligan-Topanou et al. 2015; PDC_3 Gogounou 2017). More than 50% of the nearly 100 local tractors are currently non-functional because of a lack of local expertise for repair and maintenance and difficulties in accessing spare parts. There has been an increase in provision of ploughing services for 25,000–30,000 XOF (US$50–60) by providers from other parts of northern Benin or neighbouring countries, particularly Nigeria, which has improved the access and use of tractors by smallholder farmers in Gogounou.

Table 4.1 Inventory of farm equipment in Gogounou

Owners	Farm equipment				
	Cultivator	30-hp Tractor	55/60-hp Tractor	75-hp Tractor	Total
Individual large-scale farmers	0	3	7	3	13
Individual small-scale farmers	14	3	30	0	47
Farm Equipment User Cooperatives (CUMA)	0	9	13	7	29
Communal Farmer Union (UCP)	0	1	0	0	1
Association of Women Groups (AGF)	0	0	0	1	1
Communal Agricultural Extension Service (SCDA)	0	1	1	0	2
TOTAL	14	18	51	11	94
Non-functional equipment	14	3	15	7	39
Functional equipment	0	15	36	4	55

Legend: hp = horsepower
Source Database Agricultural Extension Service (SCDA), Gogouno

Photo 4.1 A farmer benefiting from a tractor-ploughing service in Borodarou (Gogounou)

Most interviewees in our study perceive the arrival of tractors as an opportunity for agricultural expansion, although this has also increased pressure on land. Some farmers deplore that tractors sold to them are not adapted to their soils and condemn the proliferation of wrecks of agricultural equipment as a factor of failure of the envisaged agricultural modernisation.

Herbicides as Catalyst of Human–Nature and Human–Human Relations

Besides tractors, the use of pesticides has increased in Gogounou as in all agricultural zones of northern Benin. There is an ever-increasing demand for chemicals, as shown in Table 4.2, which presents the evolution of the quantities of chemicals delivered to cotton farmers by the Gogounou local extension service from 2013 to 2017. Since cotton is the most organised agricultural sector and the cotton producers are the only

Table 4.2 Inventory of chemicals for cotton production from 2013 to 2017

Agricultural campaign	Cultivated area (ha)	Chemical fertilisers (t)		Herbicides (l)		Insecticides (l)		
		NPK SB	Urea	Systemic	Selective	Alternative products	Binary miticide	Binary aphicide
2013–2014	17,788	3556	1582	29,343	20,976	40,252	32,136	4504
2014–2015	21,079	4405	1659	58,467	38,951	40,755	44,518	6716
2015–2016	20,980	4125	803	19,996	49,999	52,795	10,786	23,582
2016–2017	28,148	4270	2211	36,394	13,560	58,087	25,000	1927

Legend: ha = hectare, t = ton, l = litre
Source Database Agricultural Extension Service (SCDA), Gogounou

ones supplied through the formal input system, it is worth noting the abundant quantities of pesticides that the other farmers mobilise in the informal market to achieve their production goals. The informal market has become the main source of supply for the majority of farmers.

With illegal imports from Nigeria and Ghana, herbicides are more available in time for farmers and at unbeatable prices also supported by credit mechanisms practised by wholesalers, semi-wholesalers and even small retailers. Even many cotton producers are no longer interested in contracting input credits from agricultural extension services in the face of a more competitive informal input sector that gives more room for manoeuvre. One litre of systemic herbicide, obtainable at the Gogounou Extension Office for 5000–8000 XOF (US$10–16) after nearly a day of transactions, is easily available on roadsides and the outskirts of all the villages of Gogounou for 2000–4000 XOF (US$4–8). A semi-wholesaler who buys at 22,000–25,000 XOF (US$44–50) from wholesalers a carton of 12 one-litre cans resells it to retailers at 30,000–36,000 XOF (US$60–72) to earn 5000–6500 XOF (US$10–13). Young unemployed graduates who succeed in selling an average of five cartons of these products each week earn an average monthly income of 100,000 XOF (US$200), almost three times the guaranteed interprofessional minimum wage of 40,000 XOF (US$80) in Benin. In a context of rising youth unemployment and underemployment rates, estimated, respectively, at 14% and 70% in 2013, where more than 150,000 young people enter the labour market each year (INSAE 2013), the informal herbicide trade has become one of the most important sources of self-employment and self-fulfilment in rural Benin. In different parts of Gogounou, it was noted that several young people return from the cities to the villages to take advantage of this economic opportunity. To this category of agripreneurs can be added other young adults, teenagers and women who find the ease of cropping with tractors and herbicides as giving higher value to the farming profession and are therefore increasingly willing to invest in this activity.

Beyond the herbicide-driven agribusiness, which is an important source of employment at grass-roots level, agricultural practices have also undergone major transformations with the advent of these chemicals (Photo 4.2). Farmers classify herbicides into three categories according to their effects on land, crops and even animals:

Photo 4.2 A farmer spraying land to be ploughed by draught oxen in Bagou (Gogounou)

(i) The burners (*kpake*) are non-selective products whose effects are associated with the ravages induced by unintentional and uncontrolled fires. They are generally glyphosate-based chemicals as written on their boxes and have trade names such as *Awura, Glyohos, Herbextra, Kalach, Sharp* and others to stress their rapid and devastating effects on weeds. *Kpake* are often used in soil preparation prior to ploughing to make land more suitable for seed reception.

(ii) The screeners (*yangatime*) are herbicides considered by local farmers as selecting their enemies in the environment where they are applied. This category covers selective herbicides used during pre-emergence to secure key crops such as cotton, maize, rice and yam from their specific pests. The intelligent weed-control function recognised for products with trade names such as *Aminoforce, Atraforce, Bic, Butaforce, Heabesta, Hervextra* and others is translated into the local names attributed to them such as *tangi* or *yangatime*, which refer to the deliberate screening and choice of weeds.

(iii) The groomers (*dame*) are chemicals perceived as compacting land texture by preventing earlier development of weeds. This category includes pre- and post-emergence herbicides used by farmers to limit weed invasion into crops at a stage that may affect yield. The *dame* concept is borrowed from the French language, in which operations of compression of land structure in the construction industry are referred to as 'damage' (grooming). The role of *dame* herbicides is also similar to that of machines in the Western context that move, grind, spray and pack snow to provide smooth runways for skier safety. Herbicides in this group such as *Atraz* and others are largely associated with crop security after seeding.

This endogenous nomenclature feeds various practices and strategies at local level. The modernity of each farmer is thus measured by the way he or she combines and takes advantage of the new technologies to increase the cropping area and improve income. The competition between local farmers is strong on who will cultivate the most land during the next crop year. Those who no longer have available cultivable land even cross the district borders to seek land to borrow or buy outside Gogounou. This is the case of Yarou, who planned to go down to Borgou region to be able to acquire more land in order to increase his current 27 ha of maize up to 127 ha. Land already under crops (*tem toko*) and land that has been held fallow or was not cultivated for a long time (*tem kpa*) are subjected to two different cultivation modalities depending on the degree of 'modernity' of the farmers involved. In all cases, cost estimates in the different production systems locally practised reveal an advantage for herbicide users over non-users. For example, use of herbicides in land preparation and weed control of one hectare of maize increases farmers' income by more than 35,000 XOF (US$70). The perception of ease of cultivation and increased gains for farmers results in an ever-increasing expansion of crop-farming land to the detriment of other land users such as pastoralists, and local protected areas are subjected to a higher degree of anthropisation (see Seidou et al. 2017). Photo 4.1 shows the evolution of areas and production volumes of main crops over 20 years in Gogounou. The main crops considered are cotton, cereals (maize, millet, rice, sorghum), roots and tubers (cassava, sweet potato, yam), leguminous (cowpea, peanut, soybean, voandzou) and vegetables (chilli, okra, sesame, squash, tomato).

Figure 4.1 reveals a considerable increase in crop area during the period considered. In 20 years, the cultivated land in Gogounou has almost quadrupled from less than 25,000 ha in 1996 to nearly 100,000 ha in 2016. This increase is mainly due to *cotton*, which is the main cash crop in the region, and *cereals*, especially maize and rice, which experienced a drastic increase in their area. Maize area has quintupled over the 20 years. Rice cropping in lowlands and floodplains has developed considerably, with an area that has quadrupled over the period. This is a major challenge for pastoralists who use the *fadama* (seasonally inundated low-lying areas) as valuable grazing resources especially in the dry season. The increase in crop area over the years has not resulted in significant improvement in yields, as shown in the corresponding curve. Labour-intensive technologies such as tractors and especially herbicides have not been instrumental in intensifying production, but have instead contributed to extensifying cropping with an unprecedented expansion of cultivated land area. Changes in vegetation cover over the last three decades give more information on the dynamics of land use in Gogounou (Fig. 4.2).

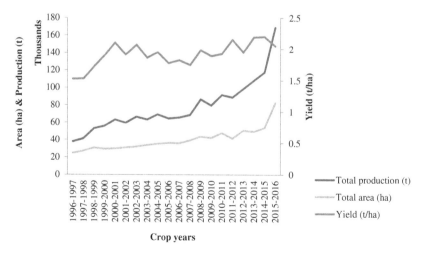

Fig. 4.1 Crop areas, production volumes and yields in Gogounou from 1996 to 2016 (*Source* Designed from the database of Gogounou Extension Service 2017)

4 Agriculture, Ecology and Economic Development ...

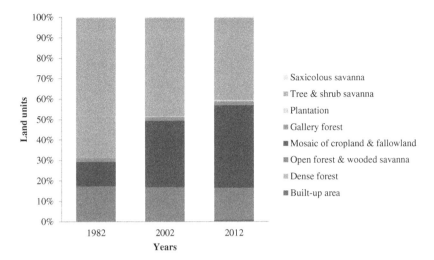

Fig. 4.2 Land units and land-cover change in Gogounou (1982–2012) (*Source* Designed from satellite images provided by CENATEL-Benin 2014)

Diachronic analysis of land-cover change in Gogounou revealed that savanna and cultivation areas are the spatial units that have undergone major changes in 30 years. Tree and shrub savanna decreased by more than 41% from 337,827 ha in 1982 to 197,893 ha in 2012, while mosaics of cropping and fallow land expanded by more than 239% from 58,754 ha in 1982 to 199,364 ha in 2012. Cultivated areas appear to have replaced the savanna, which is constantly being reduced in a context of population growth where built-up areas and private plantations continue to nibble away at common land. Anthropic pressure also on forest and woodland intensifies from year to year as crop fields are established within the two local classified forests.

Beyond favouring the expansion of crop farming, labour-saving technologies in general and herbicides in particular are at the heart of other practices connected to the increasingly harsh competition for access to resources. Herbicides help increase the power of many farmers over land. These farmers use herbicides to surreptitiously claim ownership over uncultivated spaces contiguous to their own fields. By applying these chemicals on neighbouring bush areas that allegedly host cotton pests, farmers position themselves as 'owners' of land that has not been cultivated for years. A farmer from Boro village who uses this practice stated that herbicides have thus increased his cropping capability. We also encountered cases where non-resident natives of Gogounou draw their tenure security from herbicides by having these sprayed on land they inherited from their parents.

Conflicts in relation to land expropriation have greatly increased in recent years and this does not spare the Fulani pastoralists, who are still regarded as landless foreigners, although their settlement in Gogounou dates back for several decades. Land on which they have lived for many years is claimed by those who consider themselves landowners or descendants of those who gave the land to the Fulani. Refusal to allow the crop farmers to reclaim the land could result in the poisoning of the Fulani's cattle and small ruminants. Grazing areas, livestock corridors and water resources are constantly polluted with herbicides. Such damage by pesticides to livestock-keeping is recorded not only in Benin but also in many other sub-Saharan African countries; for example, several media reports reveal cases of accidental or intentional animal poisoning in Burkina Faso (Tiegna 2008; Kambou 2013; Congo 2016).

A recent study on 100 Fulani households in Gogounou found that, in the average household herd of 25 head of cattle, 13 sheep and 11 goats, herbicides were blamed for killing two animals and causing five animals (of all species combined) to fall sick on average per month. This leads to an estimated financial cost of about 224,000 XOF (US$450) per month for the pastoralist household (Adeleke 2017). This explains why most Fulani agro-pastoralists in Gogounou, although 98% of them also use these chemicals to some extent for crop farming, perceive them as a danger to their livelihood. The Fulani consider themselves victims of this technological revolution, which has led to tons of herbicides being released into their living and livestock-keeping environment. Their cropping practices, which they consider better integrated with their livestock-keeping systems than in the case of the non-Fulani crop farmers, could do well without these products, especially since only 9% of the Fulani use tractor-ploughing services (Adeleke 2017).

The consequence is the mass departure of the Fulani pastoralists from Gogounou to seek refuge in other regions of the country or completely outside the national borders, with Togo and Ghana being the main host countries. Pastoralist civil society organisations in northern Benin revealed that 228 herds with a total of about 11,085 head of cattle and 2578 small ruminants had left Gogounou in 2012 (Boukari Bata 2012). The local agricultural extension service reported to have counted 55,000 head of cattle in the district in 2016, compared to 77,000 head of cattle in 2010 (PDC_3 Gogounou 2017). The support programme for milk and meat sectors in Benin also noted a reduction in the local livestock from 85,000 to 40,000 head of cattle because 223 Fulani settlements had left Gogounou (PAFILAV 2014). Although these figures on mass outmigration of Fulani pastoralists must be treated with caution, they nevertheless reflect the magnitude of a social fact that was hitherto underestimated but creates enormous socio-economic impact on the ground. For example, livestock market actors, mini-dairy promoters and pastoralist associations complain that the outmigration of livestock has reduced their incomes.

Herbicides also pose increasing challenges on biodiversity, as farmers associate the occurrence of several invasive species and the disappearance of some animal species with the effects of the pesticides they use. During a recent mission to Burkina Faso for backstopping country platforms in the Prolinnova (Promoting Local Innovation) network, several farmers from Sissili and Boulgou Provinces in the Centre-Eastern and Centre-Western regions, respectively, pointed to a reduction in snakes, termites and toads in their area because of the use of agricultural chemicals. Beekeepers in Burkina Faso also asked whether it would be possible to return to traditional agriculture and banish all the chemicals that are reducing the bee population and leading to lower honey harvests. Stakeholder statements suggest strong ecological and health issues related to agricultural chemicals, although these remain to be demonstrated scientifically.

Discussion and Conclusion

This study did not directly assess ecological impacts of agricultural technologies that have been widely promoted in rural Benin in recent years. However, the results obtained make it possible to draw attention to some key points.

The institutional framework for import and supply of agricultural inputs is an important factor in the analysis of the trajectories followed by these technologies. The Government's noble vision of making Benin a subregional agricultural power in the medium term and improving the contribution of the agricultural sector to the national economy should not lead to blind policies that hardly integrate the different livelihoods and ethnic groups in the rural areas. The de-liberalisation of the cotton sector and the other reforms of this kind have contributed to creating lawless spaces where private operators, unemployed youths and farmers with their respective agency play the game in realising their own agendas. Priority should be given to designing and enforcing specific regulations on pesticides, strengthening checking of chemicals released in the countryside and carrying out impact assessments for decision-making by policymakers at various levels. All this must be backed up by policies that promote better integration of cultivation and livestock keeping.

The abusive use of herbicides has major socio-economic impacts at local level, as shown in this study. Farmer-to-land relationships are strongly embedded in the dynamics of continuous expansion of fields in uncontrolled proportions that herbicides have helped to maintain. The lack of land not yet cultivated leads some farmers to be involved in expropriating land either from fellow crop farmers or from pastoralists, who constitute a competing land-use group since decades. Herbicide-based practices of territorialisation and land appropriation have completely reconfigured land-tenure security, which must be perpetually renegotiated between different stakeholders. The 'return to the village' evoked here and the increasing land transactions associated with this have contributed to deteriorating social relations, which have become increasingly tense and conflicting, although there are no statistics available to back this up. Poisoning of humans and especially livestock, and pollution of local land and water resources constitute a highly worrying dimension of the 'technological revolution' that prevails in Benin's rural areas. Livestock trade and dairying activities are affected, reducing the income of those who depend on them, including Fulani women and pastoralist civil society organisations. Even though complaints about disappearance of animal or plant species and invasion by undesired species are not yet as strong in northern Benin as in Burkina Faso, this may be only a matter of time and gradually increasing ecological awareness. In view of all the existing and potential damage caused by the unregulated 'modernisation' of cultivation, this topic must be debated in political arenas and lead to consequent decision-making and monitoring, and environmental education must follow to raise awareness of local people who are both beneficiaries and victims of the introduced technologies and the 'pesticide regime'.

Promoting community-based discussions could be a way out of the disaster pending from the way agricultural chemicals are being used. Describing and understanding what is happening with respect to the agricultural 'modernisation' is important, but finding ways to deal with or prevent its negative effects is even more important. Merely making laws and regulations might not solve the problem, because these will not be properly applied unless numerous actors within the society find it important to do so. The informal economy will always prevail and it must therefore be within the informal economy that there is enough awareness to act in a way that is responsible to the wider society. In this respect, a study by the International Livestock Research Institute (ILRI) about improving food safety in informal markets for animal-source products is very instructive (Roesel and Grace 2015). Crop farmers, pastoralists, youth, women and other actors may discover—with more awareness-raising—that many of them share similar concerns about the negative effects of chemicals on insects, wild and domestic animals and humans and could, through (possibly facilitated) discussion among themselves and also with scientists and extension service providers come to agreements on reduced and safer use of agricultural chemicals—at communal, provincial and national level. Conditions must be provided for such community-led processes.

References

Addrah, E. C. C. (1979). *Culture Attelée en République Populaire du Bénin*. Thèse de doctorat en médecine vétérinaire, No. 2. Dakar: Ecole Inter-Etats des Sciences et Medecine Vétérinaires (EISMV) de Dakar.

Adechian, S., Baco, M. N., Akponikpe, I., Imorou Toko, I., Egah, J., & Affoukou, K. (2015). Les pratiques paysannes de gestion des pesticides sur le maïs et le coton dans le bassin cotonnier du Bénin. [VertigO] *La Revue Électronique en Sciences de l'Environnement, 15*(2). https://vertigo.revues.org/16534.

Adeleke, S. L. (2017). *Effets de l'usage des herbicides et tracteurs sur les pratiques d'élevage et déterminants des conflits entre agriculteurs et éleveurs dans la commune de Gogounou. Mémoire de Master en Sociologie des Ressources Naturelles*. Parakou: Université de Parakou.

Agbohessi, P. T., Imorou Toko, I., & Kestemont, P. (2012). État des lieux de la contamination des écosystèmes aquatiques par les pesticides organochlorés dans le bassin cotonnier béninois. *Cahiers Agricultures, 21*, 46–56.

Akrich, M. (1998). Les utilisateurs, acteurs de l'innovation. *Education Permanente, 134*, 78–89.

Atachi, P. (1979). *Test de comportement sur certains herbicides CIBA-GEIGY: cas de Primagram sur maïs et sorgho, puis de Cotofor et Cotodon sur coton et arachide, en République Populaire du Bénin*. Cotonou: Ministère des Enseignements Technique et Supérieur: Rapport Etude de Pré-vulgarisation/Projet de Défense des Cultures.

Balse, M., Havard, M., Girard, P., Ferrier, C., & Guérin, T. (2015). Quand innovations technique et organisationnelle se complètent: les Coopératives d'utilisation de matériel agricole (Cuma) au Bénin. *Agronomie, Environnement et Sociétés, 5*(2), 17–23.

Boukari Bata, A. (2012). *Rapport d'activité de l'UDOPER Borgou-Alibori 2012*. Gogounou: UDOPER Borgou-Alibori.

Congo, M. (2016). *Intoxication bovine à Fada: les populations en danger*. Available at: http://www.sidwaya.bf/m-12703-intoxication-bovine-a-fada-les-populations-en-danger.html. Accessed July 15, 2017.

CTA. (2008). *L'Association Interprofessionnelle du Coton au Bénin*. Available at: http://www.inter-reseaux.org/IMG/pdf/Fiche_AIC_francais_premiere_version.pdf. Accessed November 17, 2014.

Djohy, G. (2017). *Pastoralism and socio-technological transformations in Northern Benin: Fulani innovations in pastoral migration, livelihood diversification and professional association.* Göttingen: Göttingen University Press.

Fichtner, S. (2016). La fabrique locale des statistiques scolaires. *Revue d'Anthropologie des Connaissances, 10*(2), 261–278.

Gaborel, C., & Fadoegnon, B. (1991). Le désherbage chimique du cotonnier et du maïs au Bénin: acquis et proposition de la recherche. In CIRAD-CA (Ed.), *Rapport Réunion de coordination de recherche phytosanitaire cotonnière, 26–31 janvier 1991, Ouagadougou, Burkina Faso* (pp. 135–151). Montpellier: CIRAD & CA.

Gibigaye, M. (2008). *La diffusion des innovations agricoles dans le Borgou et l'Alibori au Bénin: Cas des Coopératives d'Utilisation de Matériel Agricole (CUMA).* Thèse de Doctorat Unique. Abomey-Calavi: FLASH/Université d'Abomey-Calavi.

Gibigaye, M., Adegbidi, A., Sinsin, B. (2010). Proximité géographique et dynamique des organisations paysannes au Bénin: le cas des CUMA dans le Borgou et l'Alibori. In K. Atta & T. P. Zoungrana (Eds.), *Logiques paysannes et espaces agraires en Afrique* (pp. 67–82). Paris: Karthala.

Hughes, T. P. (1983). *Networks of power: Electrification in western society, 1880–1930.* Baltimore, MD: Johns Hopkins University Press.

INSAE. (2013). *Transition vers le Marché du Travail des Jeunes Femmes et Hommes au Bénin.* Genève: INSAE & Bureau International du Travail (BIT).

INSAE. (2016). *Principaux indicateurs socio-démographiques et économiques du Département de l'Alibori (RGPH–4, 2013).* Cotonou: DED/Institut National de la Statistique et de l'Analyse Economique (INSAE).

Jarosz, L. (2001). Feminist political ecology. In N. J. Smelser & P. B. Baltes (Eds.), *International encyclopedia of the social and behavioral sciences 8* (pp. 5472–5475). Oxford: Elsevier.

JORB. (2008). Décret n° 2007–090 portant création, attributions, composition et fonctionnement du Conseil National de la Mécanisation Agricole (CNMA). *Journal Officiel de la République du Bénin (JORB), 15*(1), 64–67.

Kambou, H. S. (2013). *Burkina Faso: Culture par les herbicides—Un phénomène aux risques énormes qui prend de l'ampleur.* Available at: http://fr.allafrica.com/stories/201308270489.html. Accessed July 15, 2017.

Kherallah, M., Minot, N., Kachule, R., Soule, B. G., & Berry, P. (2001). *Impact of agricultural market reforms on smallholder farmers in Benin and Malawi* (Final Report 1). Hohenheim: IFPRI, GTZ, & University of Hohenheim.

Kinninvo, F. (2010). Crise alimentaire: Le gouvernement béninois offre 150 tonnes de maïs au Niger. Quotidien Nokoué. Available at: http://bj.jolome.com/news/article/crise-alimentaire-le-gouvernement-beninois-offre-150-tonnes-de-mais-au-niger-800. Accessed May 10, 2015.

Kpadé, P. C., & Boinon, J. P. (2011). Dynamique des politiques cotonnières au Bénin. Une lecture par la dépendance de sentier. *Économie Rurale, 321*(1), 58–72.

Latour, B. (1986). The powers of association. *The Sociological Review, 32*(S1), 264–280.

Law, J. (2012). Technology and heterogeneous engineering: The case of Portuguese expansion. In W. E. Bijker, T. P. Hughes, & T. J. Pinch (Eds.), *The social construction of technological systems: New directions in the sociology and history of technology* (pp. 105–128). Cambridge: MIT Press.

Ligan-Topanou, O., Okou, C., & Boko, M. (2015). Durabilité agroécologique des exploitations agricoles dans la commune de Gogounou au Bénin. *Afrique Science, 11*(3), 129–137.

Liverman, D., Moran, E. F., Rindfuss, R. R., & Stern, P. C. (1998). *People and pixels: Linking remote sensing and social science*. Washington, DC: National Academy Press.

Mackay, H., & Gillespie, G. (1992). Extending the social shaping of technology approach: Ideology and appropriation. *Social Studies of Science, 22*(4), 685–716.

Maina, A. (2010). Le Bénin offre 150 tonnes de maïs blanc local au Niger. http://adammaina.blogspot.com/2010/05/le-benin-offre-150-tonnes-de-mais-blanc.html. Accessed February 15, 2015.

Marnotte, P. (1994). L'utilisation des herbicides en milieu paysan des zones soudaniennes et sahéliennes: contraintes, alternatives et perspectives. In CIRAD-CA (Ed.), *Rapport Réunion Phytosanitaire de Coordination Cultures Annuelles—Afrique Centrale, 26–29 janvier 1994, Maroua, Cameroun* (pp. 134–145). Montpellier: CIRAD-CA.

Marnotte, P. (1995). Utilisation des herbicides: contraintes et perspectives. *Agriculture et Développement, 7,* 12–21.

McCusker, B., & Weiner, D. (2003). GIS representations of nature, political ecology, and the study of land use and land cover change in South Africa. In K. S. Zimmerer & T. J. Bassett (Eds.), *Political ecology: An integrative approach to geography and environment-development studies* (pp. 201–218). New York: The Guilford Press.

Meenink, H. J. W. (2013a). Un contexte de changements institutionnels. In B. Wennink, H. Meenink, & M. Djihoun (Eds.), *La Flière Coton Tisse sa*

Toile au Bénin. Les Organisations de Producteurs Étoffent leurs Services aux Exploitations Agricoles Familiales (pp. 17–29). Amsterdam: SNV/KIT.

Meenink, H. J. W. (2013b). Faire pression pour une politique agricole et une gouvernance de la filière coton propices aux exploitations familiales. In B. Wennink, H. Meenink, & M. Djihoun (Eds.), *La Flière Coton Tisse sa Toile au Bénin. Les Organisations de Producteurs Étoffent leurs Services aux Exploitations Agricoles Familiales* (pp. 67–77). Amsterdam: SNV/KIT.

Moumouni, I. M., Baco, M. N., Tovignan, S., & Djohy, G. (2013). Appropriation of socio-technical innovation for large scale agriculture: Case study of the shared mechanization in Benin. *Schriften der Gesellschaft für Wirtschafts- und Sozialwissenschaften des Landbaues, 48*, 535–536.

Ouestaf News. (2010). *Solidarité ouest africaine: le Bénin offre 150 tonnes de maïs au Niger.* Available at: http://www.ouestaf.com/Solidarite-ouest-africaine-le-Benin-offre-150-tonnes-de-mais-au-Niger_a2987.html. Accessed February 15, 2015.

PAFILAV. (2014). *Etude des filières lait et viande. Rapport synthèse définitif, Septembre, 2014.* Cotonou: Projet d'Appui aux Filières Lait et Viande (PAFILAV)/MAEP.

PDC_1 Gogounou. (2004). *Plan de Développement Communal de Gogounou: 2005–2009.* Gogounou: Mairie de Gogounou.

PDC_2 Gogounou. (2010). *Gogounou, pôle sous-régional du commerce de bétail: Plan de Développement Communal de Gogounou, deuxième génération 2011–2015.* Gogounou: Mairie de Gogounou.

PDC_3 Gogounou. (2017). *Gogounou, pôle sous-régional du commerce de bétail: Plan de Développement Communal de Gogounou, troisième génération 2017–2021.* Gogounou: Mairie de Gogounou.

PSRSA. (2011). *Plan Stratégique de Relance du Secteur Agricole (PSRSA).* Cotonou: Ministère de l'Agriculture, de l'Elevage et de la Pêche (MAEP).

Robbins, P. (2012). *Political ecology: Critical introduction to geography* (2nd ed.). Oxford: Wiley-Blackwell.

Roesel, K., & Grace, D. (2015). *Food safety and informal markets: Animal products in Sub-Saharan Africa.* London and Nairobi: Earthscan from Routledge/International Livestock Research Institute (ILRI).

Saizonou, J. (2009). Quand l'Etat motorise des exploitations agricoles… *Grain de Sel, 48,* 28–29.

Seidou, A. A., Agbayigbo, A. A., Traore, I. A., & Houinato, M. (2017). Spatio-temporal dynamics of natural rangelands exploited by transhumance cattle herds in the Classified Forest of Upper Alibori, northern Benin. *American*

Scientific Research Journal for Engineering, Technology and Sciences, 33(1), 111–123.

Spier, R. (1970). *From the hand of man: Primitive and preindustrial technologies.* Boston: Houghton-Mifflin.

Tiegna, M. (2008). *Pollution environnementale: 23 bœufs morts empoisonnés à Banankélédaga.* Available at: http://lefaso.net/spip.php?article26448. Accessed July 15, 2017.

Vinck, D. (2012). Manières de penser l'innovation. In B. Miège & D. Vinck (Eds.), *Les Masques de la Convergence: Enquêtes sur Sciences, Industries et Aménagements* (pp. 125–148). Paris: Éditions des Archives Contemporaines.

Watts, M. J. (2000). Political ecology. In E. Sheppard & T. J. Barnes (Eds.), *A companion to economic geography* (pp. 257–274). Oxford: Blackwell.

5

Is Growing Urban-Based Ecotourism Good News for the Rural Poor and Biodiversity Conservation? A Case Study of Mikumi, Tanzania

Stig Jensen

Introduction

The aim of this chapter is to present and reflect upon the implications of urban-based ecotourism for nature conservation and rural development, building on ongoing research and preliminary findings from Tanzania. These new and intensified urban–rural activities around ecotourism are notable in Tanzania, because urban-based ecotourism can be seen as an example of a new frontier in the country. A 'frontier' is seen here according to M. B. Rasmussen and C. Lund (2018, 391) as a contact zone or a social space making the difference between 'civilization' and 'the wild'.

I will use Mikumi as the rural case study—a 3200 km² National Park, close to Morogoro, in central Tanzania. Currently, Tanzania is one of the prime ecotourism destinations in Africa, and the tourists until

S. Jensen (✉)
Centre of African Studies (CAS), University of Copenhagen, Copenhagen, Denmark
e-mail: sti@teol.ku.dk

recently consisted almost exclusively of visitors from wealthy countries in the Northern Hemisphere ('North–South' tourism). But in recent years, a growing number of urban-based Tanzanians have also shown interests in ecotourism activities in the Mikumi area.

Debates and academic literature on rural–urban relations over the past decades have almost entirely been on flows from rural to urban areas. Urban-based ecotourism is new in Africa, outside South Africa and a few other places in Southern Africa, and in most African countries, urban-based ecotourism is either non-existent or marginally developed. Nonetheless, my experience is that urban-based ecotourism is on the rise in the Global South, most notably in Asia, particularly China and India. The two main arguments regarding research activities and the relevance of looking at urban-based ecotourism in an Africa context are: (a) the increasing demand for recreational activity by a growing financially privileged class in Africa's urban areas, often labeled the middle class, which has intensified urban-based activities into rural areas connected with increasing interests in ecotourism-related activities in rural areas, and (b) growing urban-based ecotourism into rural areas has implications on both local communities and biodiversity.

This chapter is based on research findings and reflections related to urban-based ecotourism and its implications in a rural Tanzanian context and will be structured around the following three questions: Firstly, why is urban-based interest for ecotourism to rural areas growing? The aim of answering the above question is to locate and analyze the demand side of urban-based ecotourism in Tanzania. Secondly, what are the possible implications for local communities related to increasing urban-based interests? Here we try to give insight into the implications of urban-based ecotourism for rural contexts of life and livelihoods. Thirdly, we consider what implications does urban-based ecotourism have on biodiversity conservation in Mikumi? Answering this question can shed light on local environmental implications, specifically biodiversity conservation, linked to growing urban-based ecotourism. Before answering the above questions, a short introduction to selected methodological issues and other considerations related to this research will be presented.

Framing the Field

The focus in the following will be on three issues that are dealt with to varying extent.

The data available for this chapter are based on work in progress. The material has been collected over several years and during numerous field trips in Tanzania, most recently in 2015. The data collection strategy was based on mixed methods: a combination of observations, interviews, and study of written material collected from key actors and other relevant sources.

Considerations and decisions related the case choice will be highlighted in the following. After considering whether a single case or rather multiple examples would be adequate to study, it was decided to focus on a single rural case that was linked to a clear urban setting. One of the reasons for that was that insight in the urban connection is obviously needed for understanding the urban drivers in urban-based ecotourism. The prime focus was on the rural case and on studying the emerging implications for so-called local development and biodiversity conservation in the frontier, specific rural setting of Mikumi. Dar es Salaam was selected as the urban study area, despite Arusha being Tanzania's ecotourism capital for decades. Arusha is the entrance to Tanzania's, and some of Africa's, most important safari destinations: Serengeti and Ngorongoro. But Arusha had limited relevance in this context because urban-based ecotourism there is poorly developed and tourist activities in Arusha are almost exclusively linked to North–South tourism. In Dar es Salaam, the situation is fundamentally different. Dar is a megacity with diversified tourism activities both to and from the city. It is both a tourist destination and a tourist hub, primarily for North–South tourism. Furthermore, residents and particularly the expanding middle class in Dar es Salaam have recently shown growing interest in recreational activities outside the city. This could be identified as the first wave of Dar's urban-based ecotourism, consisting mainly of excursions to nearby beaches and the surrounding small islands, plus Zanzibar. Recently, a growing emphasis has been noted on ecotourism to rural areas further away. This can be seen as a 'second wave' of urban-based ecotourism. Mikumi has been chosen as the rural case for mainly three reasons: (a) There are rapidly increasing urban-based ecotourism activities in the Mikumi National Park area; (b) Mikumi is a rural area characterized by widespread poverty and according to Vedeld et al. (2012, 20), the average income in Mikumi area is only around 0.45 US$ per person per day, and therefore relevant for studying developmental implications in the local context; and (c) Mikumi National Park is one of Tanzania's largest protected areas and therefore important for biodiversity conservation.

No doubt the Mikumi case fits in a wider global trend of growth of ecotourism. Specifically, urban-based ecotourism to rural areas, as mentioned previously, has shown a remarkable upsurge and probably this will continue to intensify in the Global South. The phenomenon first requires a definition. What is 'ecotourism'? Over the years, it has been defined in several ways. Fennell (2001) has even identified 85 definitions. One of the first definitions is that of Valentine (1993, 108–109): 'ecotourism is restricted to that kind-of tourism which is:—based on relatively undisturbed natural areas;—non-damaging, non-degrading, ecologically sustainable;—a direct contributor to the continued protection and management of the natural areas used, and—subject to an adequate and appropriate management regime.' The current definition of The International Ecotourism Society (TIES) is: ecotourism is 'responsible travel to natural areas that conserves the environment, sustains the well-being of the local people, and involves interpretation and education' (TIES 2015, online).

While all these elements are important to understand what ecotourism is, in this chapter we will simply start with the baseline common denominator definition by Erlet Carter (2006) of ecotourism as tourism being nature-based. Also, in tourism research, several typologies are used, such as distinctions between 'mass tourism' and 'alternative tourism,' or North–South and South–South tourism. In this context, urban-based ecotourism is seen as a sub-category within South–South tourism, geared to leisure time spent 'in nature.'

Secondly, there are uncertainties about the socioeconomic and environmental impacts of ecotourism in the Global South. What can the possible significance of ecotourism be in the Global South context? Two important elements are the following: the rapid growth in numbers and scope, and its contribution to a country's 'economic growth' gross domestic product (GDP) and employment. Coria and Calfucura (2011) documented annual growth rates in the Global South of 10–12% per year: three times faster than the tourism industry as a whole (which is considered one of the world's largest industries). In the past two decades alone, the growth of tourism has been enormous, with 52% of these being for recreational and leisure tourism (Karis et al. 2013, 14). According to the *African Tourism Monitor* (2015: 26), Africa was and is one of the fastest growing tourism regions over recent years and one of the most important segments is 'wildlife watching tourism.' In Tanzania, the same pattern can be seen, and according to Vedeld et al. (2012, 21), Tanzania in 2011 had 1 million international visiting tourists (10 times that of 1990). This figure was kept up in 2016, with a total of 1.28 million tourist visits (*Tanzania Invest* 2017). The contribution of tourism to GDP was 17.5% in 2006 (according to Gereta 2010), with the same percentage in 2016 (*Tanzania Invest* 2017), with 25% of its foreign currency earnings. A majority of the tourists visiting can be categorized as ecotourists. The *African Tourism Monitor* (2015, 6) estimated that in 2004 in Africa direct travel and tourism employment totaled 8.7 million people. For Tanzania, it provided employment (direct, indirect, induced) for an estimated 1,337,000 people in 2014 (*African Tourism Monitor* 2015, 9), and these figures remained largely the same in recent years.

While there seems to be consensus about such 'development trends' in ecotourism, growing disagreement about the full implications related to ecotourism in the Global South has arisen. The following reflections will focus on selected positions in the literature on socioeconomic and environmental implications in the South. The literature on ecotourism's socioeconomic impact is rich, with a vast range of perspectives and contradictory views. A position articulated by Duttagupta (2012) is that tourism is seen as an effective tool in 'fighting poverty.' That position is challenged by Chok et al. (2008), who stated that: '…tourism is too often regarded as an economic, social and environmental cure-all.' The cure-all or selected benefits of tourism activities are a core area of interest in ecotourism debates. Another entry-point in the ecotourism literature is on the actors or 'stakeholders' involved. Erlet Carter (2006) focuses on benefits for the private sector, stating that nature-based tourism is one of the world's most lucrative niche markets. Power relations between different actors or stakeholders or those affected by ecotourism have been articulated by several such authors as Coria and Calfucura (2011), contending that ecotourism adds new elements to 'frontier resource conflicts': Tourism operators and indigenous communities come to compete against each other for access to resources. Along the same lines, Scheyvens (2009) has questioned the implications of ecotourism on the poorest groups in society in the following way: 'Can the interest of the poorest members of a society really be served by promoting expansion of a global industry that is founded on inequalities, where individual businesses strive to meet the interest of the market, not the poor, and where elites often capture the majority of benefits of any development, which does occur?' (ibid., 2009, 195).

The above illustrates that classical power struggles and dynamics can also be found in ecotourism where the private sector is often a powerful stakeholder with a specific agenda, as Scheyvens (2009, 193) suggested: the motivation for investors to work with ecotourism is to make profit, not to serve the poor. Another related issue discussed is ecotourism's supposed 'trickle-down effects,' and Ashley et al. (2001) argued that even if richer people benefit more than poor, also the poor still benefit, so that this form of tourism can be classified as pro-poor. Whereas literature on tourism and ecotourism's socioeconomic implications in the Global South are growing and rich, studies on environmental effects, and particularly effects on biodiversity conservation related to ecotourism, are still scarce. Often, implicitly recreational activities are seen as positive or with limited negative implications on biodiversity. As stated by Van der Duim and Caalders (2002, 743): 'Tourism has long been considered a "clean industry", without any negative effects on the environment worthy of mention.' Based on my own experience with ecotourism worldwide, it can, however, be problematic for biodiversity conservation and have severe implications. One example of potentially negative impacts of ecotourism is Tiger-tourism in the Indian National Parks. In 2012, the Indian Supreme Court decided to close down all tourist activities in National Parks due to a growing concern about ecotourism's negative impact on tigers and other biodiversity features. When the Indian National Parks were reopened, stricter regulations on Tiger-tourism were imposed in order to improve the protection of tigers and enhance biodiversity. The Tiger-tourism example might be unusual, but there is growing concern among conservationists on ecotourism's potentially negative impact on biodiversity (for more on this, see Van der Duim and Caalders 2002). This is of course a paradox, because ecotourism was also meant to 'respect the environment' and leave a minimal ecological footprint.

Analysis

In the following three subsections, I focus on answering the above-mentioned three sub-questions (cf. p. 2), starting with the reasons for the growth of on urban-based ecotourism in Tanzania. There is clearly a new demand-driven need for ecotourism to rural areas (and tourist experiences in general) on the side of urban populations, notably those in Dar es Salaam. The recent trend of urban-based tourism in Tanzania must be understood in relation to a global trend, where economic growth has improved economic living conditions for large segments of city populations, often labeled the 'new' middle classes. Among the urban middle class and/or economic elites in Dar es Salaam, three main types or categories of interests linked to ecotourism can be identified: regular tourists, investors, and tour operators. Although these are distinct categories, they are also fluid, as an investor can also be a tourist and vice versa with the other categories. The following is a short presentation of these types and their interest in ecotourism to rural areas.

Urban middle-class tourists enjoy new wealth and also for social prestige reasons experience a growing demand for recreational activities away from the city. These out-of-the-city recreational activities are carried out mainly during weekends and/or holidays, where urban-based recreational activities from Dar es Salaam to rural areas are still a relatively new phenomenon. Traditionally, non-work-related excursions outside towns and cities to rural areas almost exclusively focused on family-related activities, such as attending weddings and funerals.

The new wave of urban-based ecotourism consists of interests to rural destinations with the purpose of recreational opportunities, particularly to areas without any prior social relationship obligations.

These new urban-based ecotourists are not a homogeneous group, although the majority of them mentioned that the prime motivation is simply 'getting away from the busy and noisy city' and getting out into the countryside. Some mentioned: relaxation with the family in peaceful surroundings, others mentioned special interests such as birdwatching, mammal watching, recreational hunting, and photo safari. Many of the new urban-based tourists stated that experiences with or in nature had top priority.

Within the tourist segment, a majority said that they had already visited Mikumi and enjoyed it. Several of those articulated that they would like to come back and if possible stay longer. Others mentioned that the relatively short distance from Dar es Salaam makes it an attractive destination. Among urban-based tourists in Dar es Salaam who have yet to visit Mikumi, several mentioned that they wished to visit it primarily based on good reviews from other tourists.

The investors: Characteristic of the investors is their interest in ecotourism as an investment opportunity with possibilities for profit, in line with the position articulated by Scheyvens (2009). The investors see potential in investment projects related to urban-based ecotourism. Several of these investors had experiences with previous investments in the tourism sector, particularly in projects on Zanzibar and/or beach areas in or near Dar es Salaam. The majority of investors saw new investment potential in the growing urban-based ecotourism to rural areas as primarily due to three factors: the growth of urban-based tourism and expectations that it will continue to expand; urban-based tourism is not as unpredictable and possesses less uncertainty compared to North–South tourism, which fluctuates according to several sensitive factors, such as fear of terrorism, diseases such as Ebola, and political, ethnic, and religious unrest; and investment projects related to urban-based tourism are seen as manageable because they are often small-scale and with limited economic risks. Investment projects vary from urban-based investors being involved in joint venture projects with international, national, or local partners, or in other cases including tour operators as partners. Among investors, the

Mikumi area is articulated as attractive and several potential projects are mentioned, particularly investments in further development of accommodation facilities. Some investors also mentioned possibilities of building upon joint ventures, which already have established networks and could be further expanded. For the Dar es Salaam-based investors, ecotourism is, as Carter (2006) states, a lucrative niche market.

Tour operators are the third category. They are Tanzania-based companies, as opposed to international tour operators. Local tour operators are key providers for the urban-based ecotourists, and most of these tour operators are more or less permanently based in Dar es Salaam. Tour operators are also a heterogeneous category, ranging from full-time tour operators to ad hoc operators with other income-generating activities outside the tourism sector. Several of the Dar es Salaam-based tour operators have a rural background and can be seen as a product of local and individual financial trickle-down from North–South tourism (cf. Ashley et al. (2001) on ecotourism's implications). These tour operators have often started as local and rural-based tourist guides or drivers. This illustrates ecotourism's potential for individuals and opportunities for social and economic ascent. Several of the tour operators stated that moving to Dar es Salaam was essential for further development of the business, thus getting closer to the clients. A majority of operators articulated the importance of maintaining connections to rural areas. This category of stakeholders can be characterized as entrepreneurial people constantly focused on business opportunities and thereby responsive to both tourists and investors. A number of tour operators also offer products to North–South tourist, at times through collaboration with international tour operators.

A majority of the tour operators have experienced growth in demands from tourists visiting and wanting to visit Mikumi and expected the upward trend to continue. Several tour operators were looking to new activities in or nearby Mikumi for further developing products for the urban-based tourists. None of the tour operators would reveal information about these new initiatives as they were working in a quite competitive environment.

The area of Mikumi is clearly attractive to stakeholders and to sum up the demand side for urban-based ecotourism to Mikumi, and the following three aspects by the stakeholders can be highlighted: good infrastructure, which makes it possible and relatively easy to visit the area; easy access to nature, particularly Mikumi National Park; and access to good, reasonably priced and varied accommodation.

Implications for Local Communities Related to Growing Urban-Based Interests

To present a complete overview of local implications related to urban-based tourism is impossible, mainly due to the fact that people in Mikumi are spread out across a huge geographical area adjacent Mikumi National Park, were heterogeneous, and could not all be approached. The local heterogeneity was related to a number of factors, such as livelihood strategies and relations to land and people. Due to these and other factors, my field research activities and main data collecting were done in Mikumi town, because it is the center for ecotourism activities to Mikumi National Park, close to the entrance plus the location for most of the local accommodation facilities. One of the downsides of research based in and focused on Mikumi town was limited access to the more remote areas and to the majority of the area's most marginalized and poorest people. In order to approach and structure the data collection on local implications related to urban-based ecotourism, inspiration has been found in Sudip Duttagupta's (2012) list of the following categories of benefits related to tourism:

- Tourism is labor intensive, therefore has the ability to employ sizeable percentages of the population.
- Tourism can also promote gender equality through the employment of women in the service sector and in the informal sector.
- Tourism can facilitate micro-entrepreneurship through the formal or informal economies.
- Tourism can lead to infrastructure developments in terms of improved roadways, public transport systems, water supply, electricity supply, etc.
- Tourism allows the poor to leverage natural resources.

The following section is structured using those five abovementioned issues as the basis for discussion. Duttagupta's perception that tourism is labor intensive, therefore has the ability to employ sizeable percentages of the population, seems to be confirmed by Mikumi, for both my observations and the predominant views of interviewees document a sizeable percentage of the local population either directly or indirectly affected by tourism. Nevertheless, the basis of findings in Mikumi needs a critical reflection on jobs and job conditions in the tourism sector. Firstly, the expanding tourist activities can not only be attributed to the growth in urban-based ecotourism, but to a combination of growth in overall travel-related activities: North–South and urban-based ecotourism as well as transit travelers. Even though there is no available statistical material, it seems that transit travelers' numbers, in line with tourism to Mikumi, are growing. Overall, this results in an increased demand for various services in Mikumi. Although the tourism sector is growing, it does not always necessarily mean more local jobs. My experiences with North–South tourism in eastern and southern Africa have shown that ecotourism does not solely result in the creation of *local* jobs. The tourism sector has traditionally imported labor, especially skilled staff for servicing tourists in connection with rural tourist destinations, and imports of staff reduce opportunities for local jobs. Even though the picture in Mikumi is far from clear, it seems, however, that the majority of the new jobs are staffed by locals. An explanation could be that the recent tourism expansion in Mikumi is being caused by transit travelers and urban-based ecotourism. These types of tourists are less service-demanding, which on the one hand result in fewer jobs in middle and higher management. The jobs can, to a significant degree, be handled by locals. Furthermore, material for construction work on tourist facilities in Mikumi can most often be procured locally. Small-scale construction activities are better adapted to local capabilities. Obviously, a large proportion of jobs and materials for construction being available locally provide local opportunities for businesses and jobs/jobseekers. The volume of jobs in connection with urban-based ecotourism is another aspect of Duttagupta's (2012) picture of tourism as labor intensive, but the field material shows that there are significantly fewer jobs in transit and urban-based ecotourism. In contrast, the inflow of especially transit travelers, and to some extent urban-based ecotourists, occurs more evenly distributed over the year, whereas the North–South tourists are more seasonal. Even though urban-based ecotourism is spread over the year, visitor numbers are highest during weekends and school holidays. Regarding employment, conditions and the types of jobs in connection with the tourism sector show that the major changes can be found in the variety of jobs and length of employment. The material from Mikumi shows that demand has increased and created more diversified types of jobs. Also, a tendency is seen toward long-term employment as a result of more intense activity in the construction sector as well opportunities for staff affiliated with the accommodation facilities and restaurants. The increased activity and spread of tourists have thus provided the possibility in the direction of more permanent employment within the tourism sector in Mikumi.

As mentioned above, urban-based tourists have different requirements than the international tourists. Regarding the qualifications needed in connection with jobs related to urban-based ecotourism, two main findings should be emphasized. Firstly, as to level and form of servicing related to accommodation facilities, the international tourists expect staff to speak international languages, most commonly English. Moreover, their accommodation facilities should live up to an international level. Local tourists, however, do not have the same requirements, and language skill is not an issue. Secondly, requirements regarding food: Where a majority of North–South tourists expect 'western food' and often with products not always produced locally. Domestic, urban-based ecotourists often eat locally produced products. Overall, the growing urban-based ecotourism provides better opportunities for being accommodated locally.

One of the key findings is the importance of local networks in the recruitment of local staff in the tourism sector. The local political-economic elite plays a central role in Mikumi when it comes to recruiting staff for the sector, whereby local patron–client relations are strengthened in connection with expanding tourism in Mikumi. The reason for the local elite's stronger position is external actors in tourism, such as investors and tourist operators that use the local networks to cope with local bureaucracy (for acquiring land and other aspects related to business activities) in Mikumi. The local elite and the external actors have some common interests. The external actors need connections to the local networks for optimization of their business. Local cooperation with external actors consolidates local powers. One of the implications for the local community in Mikumi is that job opportunities depend on and are distributed to those with 'good connections' to the local political-economic elite.

Even though Duttagupta's (2012) remarks about positive spin-off in the form of jobs related to the growing urban-based ecotourism can be confirmed, and the material from Mikumi shows increased job creation, urban-based ecotourism is not as labor intensive as North–South tourism. On the other hand, there seem to be better opportunities for local tourism-related jobs to become permanent, since urban-based ecotourism takes place throughout the year. Finally, increased urban-based ecotourism contributes to strengthening local center–periphery relations, because the local elite plays a central role in the recruitment processes of staff for tourism-related activities. Thus, a form of selective local trickle-down effect linked to the powerful local political-economic elite is visible.

Gender Equality

Another conclusion of Duttagupta (2012) is that: 'Tourism can also promote gender equality through the employment of women in the service sector and in the informal sector.' A major problem in the data collection related to gender equality was the difficulties in assessing and setting up interviews with women in Mikumi. Therefore, my material is gender biased, but field observations can be used to present and reflect on some of the gender issues linked to tourism-related jobs in Mikumi. One observation is that jobs in the tourism sector are gender-skewed. For example, many jobs in connection with lodging are filled by women, the same for functions such as receptionist or cleaner, while men mainly undertake more physically demanding jobs in construction, security tasks, and tour guide work. But intensified tourism in Mikumi did increase job opportunities also for women. Paid jobs created the possibility for an independent financial basis for more people. In addition, the growth in tourism means that local people meet tourists. This also applies to urban-based ecotourists practicing a modern lifestyle. Increased exposure to modern lifestyles and ideas has implications locally, because tourists are perceived as successful and thus become lifestyle role models. Particularly, the urban-based ecotourists are role models for young people in Mikumi.

Overall, it is difficult to give a clear response to Duttagupta's remarks. Primarily due to the abovementioned limitations in my material, it seems that increased tourism may promote gender equality: There *are* more jobs for women and thereby increased opportunities for establishing their own financial position and some economic independence. Secondly, other factors such as exposure to and direct confrontation with a 'modern' and western-inspired lifestyle could question the traditional lifestyles, including gender roles.

Micro-Entrepreneurship

The following section will address Duttagupta's remarks on whether '….tourism can facilitate micro-entrepreneurship through the formal or informal economies.' Micro-entrepreneurship is not a new phenomenon in the Mikumi area, according to local sources. There is a general consensus in Mikumi that micro-entrepreneurship has played and still plays a central role in developing and branding Mikumi as a tourist destination. The following three key dimensions can be drawn from my material. Firstly, liberalization of the economy in Tanzania has created space and better opportunities for private initiatives and improved conditions for micro-entrepreneurship in Mikumi. Secondly, the new young generation sees opportunities in micro-entrepreneurship activities as possible foundation for a better life different and more modern than their parents' lives. Therefore, the predominance of young people among the new micro-entrepreneurs related to ecotourism and other new initiatives, locally. When tourism is attractive to the entrepreneurial young, it is both due to earnings possibilities and because tourists for many young people represent role models. Thirdly, urban-based ecotourism to Mikumi has indeed created a number of new earning opportunities for people in the area. Three of the most visible micro-entrepreneur activities related to ecotourism are: guide activities, producing and selling handicrafts, and sale of local products to take home for consumption. Urban-based ecotourism has thus created opportunities for micro-entrepreneurship development in connection with diversification of the product range, for example, looking for products for urban ecotourists other than the ones for North–South tourists. Even though the homogenization of tourists within the following categories is not unproblematic, many of the urban-based tourists focus on enjoyable activities in nature, such as a picnic in the wild, and to be out in a peaceful, beautiful landscape. Another example of local micro-entrepreneurship is the establishment of new restaurants with menus primarily directed at the domestic tourist segment. The same applies to the expansion of accommodation facilities, with both broad spectra as well as a specialized approach to the market. A third type of micro-entrepreneurship focusing on take-home products has also developed in recent years. The target group for these products is not exclusively urban-based ecotourists, but also transit travelers, The 'product range' is locally produced goods such as charcoal, different items made from wood or sisal, vegetables, and fruits. The sale of these takes place along the main road, often with limited sales facilities. A local network has been established in Mikumi between the people selling and the manufactures, and the charcoal is made by the poorer people in Mikumi.

In summary, Duttagupta's statement can be confirmed that tourism contributes to micro-entrepreneurship. A combination of factors has promoted opportunities for local micro-entrepreneurship to flourish, whereby increased demand, enterprising young people and national policies are the most central elements.

Infrastructural Upgrading

Regarding Duttagupta's remark that: 'Tourism can lead to infrastructure developments in terms of improved roadways, public transport systems, water supply, electricity supply, etc.,' in the Mikumi area, infrastructure directly benefitting the locals has not noticeably been improved. Even though several infrastructural upgrades in Mikumi can be identified, these are almost entirely for tourists, for example, in hotels, with improvements of water, sanitation, access roads to accommodation, electricity, and Internet access. Even though some of the local employees in the tourist industry also have gained access, the overall effects on local society seem to be limited.

Tourism Allows the Poor to Leverage Natural Resources

An additional statement of Duttagupta on the effects of tourism is that it '… allows the poor to leverage natural resources.' On this, I report the following two key findings. Firstly, for those who live close to the Mikumi National Park, the costs are higher than the benefits, primarily because intruding wildlife primarily from Mikumi National Park makes it difficult for any agriculture production on the land. This is due to several factors. One is that wild animals enter, eat, or in other ways negatively harm crops and livestock. Another reason is that locals are concerned about the lack of support from the authorities and especially from the local National Park representatives, as they are seen as only concerned with protecting wildlife. The local views on Park staff are in accordance with interviews done by Newmark et al. (1993, 77) with people living near protected areas, where '71% held negative or neutral attitudes towards protected areas employees.' Additionally, growing local dissatisfaction and demands for compensation mechanisms for damage caused by wildlife prevail. A combination of lack of compensation schemes and conservation-oriented National Park staff causes frustration locally. People living near Mikumi National Park do not feel that nature conservation and tourism improve the possibilities for exploiting nature: They find it to be quite the opposite. This is along the same lines as previous studies (for more on this, see Brockington and Igoe 2006; Vedeld et al. 2012) and constitutes a danger for ecotourism and environmental conservation in the long run.

Among residents living further away from Mikumi National Park, the growing urban-based ecotourism did create new possibilities through utilizing the nearby National Park and earnings for selling items to urban-based tourists and travelers. The improvement is attributable to increasing demand, particularly for charcoal. The charcoal is produced locally and often by the poorest segment in the local community, which has usually been marginalized. They have grasped the new opportunities, although this is done via more intensive use, or rather exploitation, of nature. Also, this endangers long-term conservation aims.

What Implications Does Urban-Based Ecotourism Have for Biodiversity Conservation in Mikumi?

The protection of biodiversity has a nominally high priority in Tanzania, and large parts of the country have been designated as National Parks, including Mikumi. The aim of these parks is to conserve and protect biodiversity against human exploitative activities. Below, I present initial data about the scale of nature conservation, both nationally and locally, and the role and significance of urban-based tourism for this. The Mikumi case is again the point of reference. As the data are limited, I will offer some general reflections, especially in the last part about a possible scenario with future implications for Tanzania's policy of protection of biodiversity in light of the growing urban-based ecotourism.

Nature conservation regulations in Tanzania are very comprehensive, and according to Vedeld et al. (2012, 21), 24% of Tanzania's total land is set aside as protected areas, of which 17.4% comprises 15 National Parks and 34 game reserves. Tanzania is one of the countries with the largest proportion of the country designated as nature reserves for protection of biodiversity. In Mikumi, the National Park covers 3230 km^2 and is the fourth largest protected area in Tanzania. The larger part of Mikumi National Park is closed off and closed for exploitation. A key finding regarding biodiversity conservation is the striking difference in biodiversity management within and outside the National Park. The authorities' approach to the management of biodiversity in protected areas is based on a classic fence-and-fine approach. The authorities' interest in the exploitation of nature, including biodiversity, *outside* Mikumi National Park, however, is non-existent. Several people stress that areas outside the Park are severely degraded, both the local biodiversity conditions as well as the scenic surroundings. Regarding the possible implications of urban-based ecotourist activities on biodiversity in Mikumi National Park, there seems to be a consensus that these are marginal and insignificant, primarily because the area that urban-based ecotourists and other tourists visit in Mikumi National Park is extremely small; the National Park's biodiversity as a whole has undergone limited effects as yet.

However, within the tourism sector of Mikumi National Park, the urban-based ecotourist activities do have implications on wildlife and other tourists, firstly, because these new ecotourists have strongly increased numbers. Several local sources mention that the growing activities in the National Park have affected some animal species. For instance, Lions has been much easier to spot in the past. Whether this means a decline in numbers or different, more reclusive behavior of the animals is difficult to say. The same sources state that there is no immediate concern for the lion population in Mikumi National Park.

Another aspect about urban-based ecotourists, which does not directly affect biodiversity, is that respondents articulated concerns about the increased tourism and in particular the behavior of urban-based ecotourists, which is different from North–South tourists. For some North–South tourists, Mikumi National Park is no longer an attractive destination, due to what could be labeled a 'zooification.' Even though urban-based ecotourists are not a homogeneous group, several sources state that some within this group see a visit to the National Park as a cozy excursion and exhibit noisy behavior similar to that in a zoo. This behavior affects the experience of the foreign (North–South) game tourists in Mikumi National Park, who get disenchanted. The majority of the urban-based ecotourists primarily focus on Mikumi National Park's recreational qualities and contrary to the majority of North–South tourists who focus on experiences with wildlife. In other words, these categories of tourists have different views on Mikumi National Park and its qualities. A predominant part of the urban-based ecotourists focuses on nature from a landscape and recreational perspective, whereas the North–South tourists focus more on the natural environment and biodiversity in a more selective species perspective.

In light of the recent developments in Mikumi with increasing urban-based ecotourism and growing dissatisfaction from North to South tourists, the following reflections on recent trends' possible implications for the biodiversity conservation in the future can be offered. The growing dissatisfaction among North–South tourists is important, but not the most important aspect for conservation of biodiversity in the future. However, the growing urban populations' interest in nature, as evidenced in the growth of domestic city-based ecotourism, could have an impact on the conservation of biodiversity in the future, both locally and more generally in Tanzania, and alter the discourse on nature conservation and biodiversity. Traditionally, the fight for nature conservation in Tanzania has been driven exclusively by international actors, in alliance with the Tanzanian state as the implementing agency. The international interest for nature conservation in Tanzania can be traced back to colonial times and the former colonial power's extensive intervention and regulation, which laid the foundation for today's protected areas and parks. In post-colonial Tanzania, environmental policies are still based on an alliance between the state and international conservation nongovernmental organizations (NGOs). As Vedeld et al. (2012, 20) states, a major contribution to the global policy architecture on nature conservation was steered by lobby groups such as the World Wildlife Fund (WWF), Conservation International (CI), the World Conservation Society (WCS), and African Wildlife Foundation (AWF). The alliance between these INGOs and the state has continued colonial regulation and expanded the area set aside as protected areas. According to J. Bluwstein and J. F. Lund (2018, 463): 'The share of Tanzania's surface area that is targeted by conservation has grown steadily over the past decades. Today estimates suggest that around 45% of Tanzania's terrestrial area is under some form of conservation-related protection.'

The regulations for protecting biodiversity were based on a Malthusian approach to people and their expanding exploitation of nature, whereby people and especially local communities were perceived as enemies of nature. The question is whether the new urban middle classes' interests in nature could open up for a Tanzanian civil society-based demand for nature conservation and whether they would share the same argument on the indeed growing clash between expanding populations and nature conservation and use.

While the primary focus among urban-based ecotourists is not specifically on the conservation of biodiversity but rather on the interest in 'nature without people,' there could be a reconfiguration of the fight for nature, where the urban middle classes link up with international nature conservation NGOs and the state in the fight for nature, seen as national wealth. Such a new alliance would undoubtedly create a stronger future foundation for the conservation of biodiversity in both Mikumi and Tanzania in general. This scenario is just one of several possibilities, but the growing urban interest in nature will under any circumstances lead to a wider nationally founded interest in nature, which was previously driven primarily by international interests and hardly sustainable for the future.

Summing up, it can be said that the growing urban-based interest in nature in Tanzania is in principle positive for the conservation of biodiversity, both in the short-term and possibly also in the future, especially when it comes to biodiversity in protected areas—depending on strong national policies. Regarding biodiversity maintenance *outside* the protected areas, the findings from the Mikumi area show that nature is significantly degraded. There are no signs that this will change: The threats to biodiversity outside protected areas will likely grow, because of expanding population, resource scarcity, and agricultural and other competition.

Conclusion

Is urban-based ecotourism is good news for the poor and for biodiversity conservation in Mikumi? The short and short-term answer from this frontier is yes. For Mikumi, the growing level of activities via urban-based tourism can be seen as expansion of capitalism resulting in new opportunities for many people in Mikumi. The three most notable aspects are that local-based tourism has created and increased job opportunities for people in Mikumi; that some of the poorest in the Mikumi area have increased opportunities for earnings, primarily through an increase in local product sales, in particular of charcoal; and that access to jobs in Mikumi's tourism industry has enlarged relations with the local political-economic elite, which gets more powerful and access to which is needed to secure jobs and related advantages. The last issue means also that a growth of tourism can contribute to social polarization in a number of areas, because some people get opportunities while others remain marginalized.

When it comes to the conservation of biodiversity in relation to growing urban-based ecotourism, it is generally positive only as far as the Park is concerned. The adverse impact is mainly seen in the non-protected areas in which exploitation has been intensified, mainly for charcoal production, whereas the effects on biodiversity in Mikumi National Park are more difficult to measure. The most important perspective for the conservation of biodiversity is the possible effects of the urban middle class growing interests and demand for nature, thereby necessitating its preservation as a 'commodity.' These new urban-based demands could have long-term implications on the fight for nature conservation in Tanzania, so far driven by international nature conservation interests and implemented by the state. Even though there seems to be short-term positive news for the poor and nature in Mikumi with increased urban-based ecotourism, it is not, as Chok et al. (2008) state, a cure-all. The benefits are segmented and uneven. Apart from the sheer weight of ever-growing numbers of tourist, visitors in the nature areas may have a long-term deleterious effect—subverting the 'ecotourism' idea, locally, tendencies toward polarization should not be underestimated, as there are people who feel marginalized, both regarding economic development and due to the regulations to protect nature. Some of them would possibly be further marginalized in the future. There have been conflicts, though still small-scale, and the most recent developments did not solve problems such as the infringement of wildlife from Mikumi National Park into agricultural lands. These and other conflicts could potentially expand and escalate if not targeted, and then we will see that, as Coria and Calfucura (2011) mentioned, that ecotourism can add '…new elements to frontier resource conflicts.'

References

African Development Bank. (2015). Unlocking Africa's tourism potential [Special issue]. *African Tourism Monitor, 3*(1).

Ashley, C., Row, D., & Goodwin, H. (2001). *Pro-poor tourism strategies: Making tourism work for the poor. A Review of Experience*. London: Overseas Development Institute, International Institute for Environment and Development and Greenwich: Centre for Responsible Tourism, University of Greenwich.

Bluwstein, J., & Lund, J. F. (2018). Territoriality by conservation in the Selous-Niassa corridor in Tanzania. *World Development 101*, 453–465. Advance online copy: https://doi.org/10.1016/j.worlddev.2016.09.010.

Brockington, D., & Igoe, J. (2006). Eviction for conservation. A global overview. *Conservation and Society, 4*(3), 424–470.

Carter, E. (2006). Ecotourism as a western construct. *Journal of Ecotourism, 5*(1–2), 23–39.

Chok, S., Macbeth, J., & Warren, C. (2008). Tourism as a tool for poverty alleviation. A critical analysis of 'pro-poor tourism' and implication for sustainability. *Current Issues in Tourism, 10*(2–3), 144–165.

Coria, J., & Calfucura, E. (2011). *Ecotourism and the development of indigenous communities: The good, the bad, and the ugly* (Working Papers in Economics, No. 489). Gothenburg: University of Gothenburg.

Duttagupta, S. (2012). Pro-poor tourism development. The case of the Endogenous Tourism Project, India. http://ecoclub.com/articles/823-120504-duttagupta. Accessed August 5, 2016.

Fennell, D. A. (2001). *Ecotourism: An introduction* (2nd ed.). London: Routledge.

Gereta, E. (2010). The role of biodiversity conservation in the development of the tourism industry in Tanzania. In E. Gareta & E. Røskaft (Eds.), *Conservation and natural resources*. Trondheim: Tapir.

Karis, A., et al. (2013). *Impact of Tourism on Wildlife Conservation*. INTOSAI Working Group on Environmental Auditing (WEGA).

Newmark, W. D., et al. (1993). Conservation attitudes of local people living adjacent to five protected areas in Tanzania. *Biological Conservation, 63*(2), 177–183.

Rasmussen, M. B., & Lund, C. (2018). Reconfiguring frontier spaces: The territorialization of resources control. *World Development, 101,* 388–399.

Scheyvens, R. (2009). Pro-poor tourism: Is there value beyond the rhetoric? *Tourism Recreation Research, 34*(2), 191–196.

Tanzania Invest. (2017). Tanzania tourist arrivals increase by 12.9% in 2016 to reach 1.28 m. Online at: http://www.tanzaniainvest.com/tourism/tourist-arrivals-reach-2016. Accessed September 20, 2017.

TIES (The International Ecotourism Society). (2015). What is ecotourism? http://www.ecotourism.org/what-is-ecotourism. Accessed September 20, 2017.

Valentine, P. S. (1993). Ecotourism and nature conservation: A definition with some recent developments in Micronesia. *Tourism Management, 24*, 107–116.

Van der Duim, R., & Caalders, J. (2002). Biodiversity and tourism—Impacts and interventions. *Annals of Tourism Research, 29*(3), 743–761.

Vedeld, P., et al. (2012). Protected areas, poverty and conflicts. A livelihood case study of Mikumi National Park, Tanzania. *Forest Policy and Economics, 21*(12), 20–31.

6

Losing the Plot: Environmental Problems and Livelihood Strife in Developing Rural Ethiopia—Suri Agropastoralism Vs. State Resource Use

Jon Abbink

Introduction

Agropastoral economies of Ethiopia cover an important land surface (ca. 40%) and produce a substantial number of items, from milk, meat, hides, skins and animals for export, next to providing livelihoods to several millions of people in often precarious areas with low rainfall. Land use in such semi-arid agropastoral areas of Ethiopia is geared to the spread and relative scarcity of resources and historically took place in relatively sustainable patterns. Today growing population pressure, competition from cultivators, and climate change/variability undermine productive conditions. The lands inhabited by agropastoralists are now also being redefined for use and resource extraction in macro-growth model policies that insufficiently connect to the existing economies and environmental conditions in place. The reasons for this are of a primary political (establishment of state authority in 'marginal areas') and economic (national export

---

J. Abbink (✉)
African Studies Centre, University of Leiden, Leiden, The Netherlands
e-mail: abbink@ascleiden.nl

© The Author(s) 2018
J. Abbink (ed.), *The Environmental Crunch in Africa*,
https://doi.org/10.1007/978-3-319-77131-1_6

development, direct land access/expropriation, new energy resource construction, such as hydro-dams) but have social and hegemonic aspects as well. Environmental–ecological considerations are not paramount in these policies, which are rather fed by perceptions of 'near-empty, poor and remote lands' with people that 'need education, employment and modernity.'

Also the Lower Omo Valley of Southwest Ethiopia, a previously 'remote area' (as seen from the political center), is the scene of massive 'development' ventures via infrastructure works, roads, and large-scale agrarian plantations for mono-crops like sugar, cotton, or biofuels. They affect virtually all local peoples (e.g., Mursi, Me'en, Suri, Kara, Kwegu, Nyangatom), many of them agropastoralists. The area is described well in a growing number of studies, and its problems are much discussed in the wider popular press and by advocacy groups (e.g., Oakland Institute 2013a, b; Perry 2015; HRW 2012, 2017; Newsome 2015; Temperley 2015; Vidal 2015). But the region's transformation marches on inexorably and is not significantly modified to address looming environmental decline, ecological imbalance, and new socioeconomic dependency of local peoples.[1]

This chapter discusses an example of one locally affected group, the Suri people, one of the 'marginal' agropastoralist groups living in this now coveted Ethiopian territory and in a double quandary: They face (a) territorial shrinkage and (b) a qualitative decline of the local ecological conditions, disturbing their economic system of cultivation, artisanal mining, gathering, and especially transhumant pastoralism. And related to this, they become more dependent on outsiders instead of on their own resources. Their environmental skills and knowledge are deemed irrelevant in models of 'modernized' large-scale commercial agriculture for export crops. The future of livelihood systems and of the people themselves is—in the view of the Ethiopian state elites—to be altered radically. I will outline this state-led modernization project—supported by World Bank money and some donor countries[2]—that has already drawn a trail of irreversible landscape alteration and which has not provided more opportunities for local people but made them, and the local ecology, more vulnerable. The Ethiopian state is an undiminished example of what J. Scott (1998), J. Markakis (2011), and others have described as a top-down central planning machine that makes blueprints and applies them with little contextual adaptation, and with citizens approached as subjects to be made legible in, and dependent on, the national project.

In this interpretive case study, based on long-term field observations in the last 15 years as well as on interviews with some administrators and local experts, I describe some key economic and environmental aspects of this expansion for the Suri people, in conjunction with political and cultural concomitants of the exercise. Some comparative notes on neighboring groups facing the same challenges will also be made. A political ecology approach forces itself upon this case study: 'nature' and 'environment' interact with humans in specific political and economic constellations impacted by power differences. The interdisciplinary political ecology approach can help to see how developmental trajectories and environmental changes are produced by a governmental politics of resource definition and appropriation that can have *direct* consequences for local ecological conditions, thereby rearranging power and dependency relations. The power dimension is essential in estimating the causalities of livelihood and environmental change in the Omo Basin, as it reveals the project of state expansion and political command structure installation in the area.

The developmental expansion of the federal Ethiopian state in marginal peripheries since ca. 2008 is substantial, and ostensibly done for the 'national benefit' and overall economic development. The wealth and profits to be generated from the new infrastructural, energy, and agrarian projects would be 'for all' and would modernize lifeways and raise standards of living. However, in this epic state venture, local peoples active in agropastoralism, agrarian cultivation, fishing and hunting-gathering livelihoods have little to offer and are relegated to a secondary position: their predominantly subsistence economies and local knowledge systems are seen as 'backward' and as 'under-exploiting' the resources of the region, not yielding profits (for the state). Basic differences as to the perception and valuation of space, place, and the natural environment have become evident, and these are rooted in the political ecology of space (cf. Korf and Schetter 2012). D. Turton already demonstrated in a seminal paper (2011) that this contestation, if not conflict, is related to both political–economic as well as sociocultural differences in approach to the 'natural environment,' the landscape and its management, and even in the definition of what 'nature' or the 'environment' *is*. As Wagstaff noted in a (2015) study, elements of cultural hegemonism are clearly visible: The Ethiopian state seems to find these people primitive and embarrassing in their way of life—e.g., their bodily culture, lack of 'literacy' and modern knowledge, specific customs of mourning, dueling, etc. Thus, the state is reproducing, in similar form, the denigrating narratives of the past (cf. Ellison 2012).

The transformation of people like the Suri agropastoralists is framed in this context of national economic development and civilizational discourse that the Ethiopian government initiated in the area. The process is perhaps another instance of the 'great transformation,' the radical change in human society famously described by Karl Polanyi in his book *The Great Transformation* (2001 [1944]), whereby a largely subsistence-oriented social economy is turned into a market-oriented commercial society based on gradual commodification, with important social, political, and environmental consequences. But in Ethiopia, it is strongly *orchestrated* along a scale of values and intrusive governance techniques defined, indeed imposed, by the state. We here look at the redefinition of 'space' and 'place' of local groups vs. the state as cultural, not only economic, concepts, and at the adverse environmental effects, suggesting gradual ecological decline and impoverishment.

A related argument of the paper will be that not only ecological preconditions for sustainability are eroding in many settings of the rural-developmental enterprise, but also that customary ethno-ecological knowledge of local peoples is receding, displaced by a discourse of direct exploitative appropriation of the environmental 'resources' that 'do not need' contextual appraisal. This tends to cause disequilibrium and in the longer-term ecological deterioration (in soil and water quality, biodiversity, and productivity). There are good reasons to expect another local 'crunch' that will undermine sustainability and ecosystem stability. In its radical development policies, as laid down in the 'Growth and Transformation Plans' (GTPs) I and II (2010–2015 and 2015–2020), the Ethiopian government, with perhaps good intentions, also seems to renege on earlier ventures to commit (with donor funding) to good environmental management, sustainability, and biodiversity policies (cf. FDRE and UNDP 2006).

In most places in Africa, including densely populated Ethiopia, there is inevitable contestation over resources and their use, and over space and place and their cultural ramifications. This subject in general has been well studied within global contexts of economic expansion, dispossession, and exploitation (Li 2010; Sassen 2014). In recent years, a real flood of new studies has highlighted these processes in Ethiopia; here also, development implies destruction and forced change (cf. Ashkenazi 2012). On Southern Ethiopia, especially the South Omo Zone, there are many studies of space and place mutations (Turton 2011; Girke 2013; Eyasu and Feyera 2010; Seyoum 2015; Tsegaye 2016; Wagstaff 2015, etc.), as well as numerous press reports on the environmental changes.[3]

Basic local perceptions of livelihood and environment or of 'nature' and ecological conditions in Southwest Ethiopia are in stark opposition to those of the new claimants to the land: the state agencies, commercial companies, and private/foreign investors.[4] Some peoples, like the Suri, living west of the Omo River and south and west of Maji town, are faced with the physical and social 'shrinkage' of space and the subversion of their homeland and orientation of place. They are not even 'localized,' i.e., put in their own small place (cf. Turton 2005, 258, 271 on the Mursi), but seem to even meet a denial of their right to *be* there, as their lands are potential state investors' territory. A similar process was visible among the Mela people east of the Omo (cf. Buffavand 2016, 2017; Stevenson and Buffavand, forthcoming).

A case study[5] on Suri illustrates current processes and policies of state-making via land-oriented sovereignty claims, buttressed by a legal framework subverting local citizens as economically and politically relevant agents, and whereby the 'ethno-ecology,' the customary interactive livelihood system of—in this case—the local agro-pastoralists as customary keepers of the land, tends to be denied or deleted.

The first part of this chapter will elaborate more on the geographical and ethnic conditions and some aspects of local perspectives of space, place, and territoriality among groups in the Ethiopian Southwest, and indicate their different modes of adaptation so far. The second part is about the 'landscape reconstruction' effected by the Ethiopian state and its developmental ventures. The third part contrasts 'ethno-ecological' views of local peoples—especially the Suri—with the perspectives brought by the state and its redefinition of this former 'periphery' both environmentally and socially. It will appear that the local 'ethno-landscape' as created by Suri (and neighboring groups) is not seen as something of value or to be preserved as managed by state-induced investors and administrators. Political ecology—here in the sense of an organized, state-led process of reassigning 'resources' and power—is seen to trump the cultural ecology that marked the local peoples and their environmental interactions based on adaptation. The experience of livelihood change and households adrift shows the problematic social and environmental reconfigurations. The final part is the conclusion, summarizing the argument and the preliminary findings.

The Lower Omo Valley

The Omo Valley is a region with great landscape and hydrological variety. It has been inhabited for several thousand years although in low densities and largely 'managed' by local peoples (cf. Gil-Romera et al. 2011). It is an area of hot plains and low hills but with cool mountains in the Dizi and Me'en areas north of the town of Maji, and in the Southwest (Mt. Naita).[6] The soils in the Suri area are mostly fluvisols with some patches of lixisols in the central area, ferralsols more to the east toward the Omo, and calcisols along the border with South Sudan (cf. FAO Soil Profile Database 1998; Belete et al. 2013, 23).[7] The mountainous parts and foothills still have montane forests and diverse plant and wildlife, although rapidly declining. The Omo River bank forests are a place of rich wildlife and biodiversity (cf. Carr 2017, 18–19, 68, 69, 95, 151). As a whole, the Omo Valley basin fulfills essential functions in a wider regional ecology and hydrology of Southwest Ethiopia up to Lake Turkana in Kenya (Avery 2012), and in 2016–2017 provides still rich, but diminishing resources for agriculture, hunting and gathering, livestock keeping, as well as beekeeping and alluvial gold panning, all traditionally carried out by the local peoples.

The southeastern parts of the Suri area, bordering Nyangatom settlements, are rather poor and semi-arid, but planned by the Ethiopian government to be filled with sugarcane plantations (Fig. 6.2) and to be irrigated with massive flows of water from the Omo. This area of mainly low bush and grass plains was used for hunting and gathering by local people and for shifting cultivation by newly settled Nyangatom after ca. 1995, but is now gradually closed off. Rainfall in the Valley ranges from over 1800 mm/year[8] in the Northern and Western parts of the basin to less than 300 mm/year near Lake Turkana. Agropastoralism—combining extensive, transhumant livestock holding, and cultivation on rotating fields, often changed—is the best economic strategy, attuned to the local ecology and based on (limited) mobility. It was practiced for ages and has allowed sustainable livelihoods and environmental continuity. Along the rivers (especially the Omo), land is (was) used for river-bank cultivation, which yielded good harvests and had great potential down along the Omo (see Eyasu et al. 2015). But it is now strongly discouraged by the government. Bodi (Mela-Chirim) and many Mursi people can now no longer reach the Omo river banks due to the sugar plantations (see below) and are seriously constrained in their resource use, even in basic cultivation.

Since the Lower Omo area became part of the Ethiopian state in the 1890s, the agropastoral production systems were never invested in; rather they were exploited (slave and serf labor power, cattle predation; cf. Garretson 1986). Also under the present EPRDF government, investments in sustainable agropastoralism and local knowledge systems, that have evolved over time, are not seen worth it, in contrast to a growing body of scientific insights showing (agro)pastoralism profitable and appropriate (e.g., Behnke and Kerven 2013; Breu et al. 2015; Krätli et al. 2015). Among the local ethnic groups in the area, the Me'en north of the Omo River have in addition to mixed subsistence agriculture developed the cash crop production of coffee, wheat, *t'eff* (Eragrostis tef) and sesame, sold to highlander traders.

The Lower Omo Valley was long considered a 'wilderness area' *par excellence*—remote, peripheral and, according both to the first Western travelers (C. Bulpett, A. Bulatovich, D. Smith, V. Bòttego) and the Ethiopian highlanders who came there in the early 1900s, inhabited by mostly 'backward and uncivilized people.' But it is now spatially redefined as a prime economic area that will bring 'export revenue' and profits due to commercial agriculture, large hydro-electricity dams, as well as gold mining, taken over from locals.

The People and the Context

The agropastoral Suri (ca. 34,000 people[9]) live in the Bench-Maji Zone of the Southern Regional State of Ethiopia, an area of savannah lowland and mountains of ca. 1800–2500 m. Their *woreda* (= district) area of some 4700 km² borders South Sudan and has a mixed altitude level, consisting of hills (10%), lower rugged areas (35%), and savanna plains (55%) with several rivers transecting the area (Tum, Kaari, Koka, and Kibish). The land shows substantial flora and fauna species diversity, is semi-arid in lowland places and vulnerable to variable rainfall, but has sufficient water in the hills. As noted above, soil fertility is mixed, but the area has good grazing lands for livestock. The lower savannah areas are mostly inhabited by Suri, Nyangatom, and by some Me'en groups (Mela, Chirim, Nyomoni),[10] while the cooler mountains and foothills are the home of Dizi (an Omotic-speaking group) and various highland Me'en groups (with rainfall of up to 2500 mm/year). But Suri traditionally also had villages in the hills of Tirma and Naita near the border with Sudan, before being chased out by force by Nyangatom in the late 1980s. In fact, Suri always preferred the cooler hills for horticulture and staple crop fields, and used the plains for livestock herding, a 'dual pattern' of settlement and eco-niche use that was in place for several hundred years, although highly mobile.

The Suri traditionally have a food-secure economy—with seasonal dips—but today face particularly acute problems of survival and conflict. They went through a period of crisis and turmoil notably in the past 30 years, due to state encroachment, regional population growth, effects of the Sudanese civil war, and growing inter-group competition due to reduction of territory. Signs of climate change effects are noted (slow drying out of the area, and water and pasture scarcity), but are not yet pervasive. Conflict is evident in clashes with neighboring groups—Nyangatom, Anywa, and Sudanese (Toposa) infiltrators (see Abbink 2009a; Wagstaff 2015)—and the state agents (government administration, agrarian investors, and army units) coming to their land. Other neighbors with whom ambivalent tensions have long existed are the Dizi (some 36,000) who are sedentary agrarian cultivators, and the shifting cultivator and mixed farmer Me'en people (ca. 155,000).[11] Especially, in conflict with Toposa and Nyangatom, numerous people were killed over the past two decades, with nefarious consequences for the social fabric of local society. Internal Suri strife has also increased notably. The Suri being a 'famous' people in the global tourist discourse (many photo books, touristic articles, and documentaries[12]) has not helped them in developing means or connections to defend their way of life.

Major changes with which local peoples in the Ethiopian Southwest have to deal since the past decade are the insertion of the new large-scale agrarian plantations on expropriated land, the damming of the Omo, irrigation schemes, coerced socioeconomic change (away from pastoralism), and mandatory villagization. A slow economic disempowerment of the local people is in progress, as their agrarian activities—such as river-bank cultivation, crop rotation, and livestock herding—are discouraged and territorially narrowed down. Different conceptions and definitions of 'nature,' 'environment,' and 'resources' compete, so as to change the *meaning* of local livelihoods and landscapes.

Some aspects of these much promising economic developmental schemes of the state—the Gibe-3 river dam, sugar plantations and factories, irrigation for large commercial farms, and settlements of workers imported from elsewhere—must therefore be seen from the other side: that of the local peoples forced to deal with and adapt to them. The sugar plantations of the Ethiopian Sugar Corporation (a state monopoly) are well described already (e.g., Keeley et al. 2013; Dessalegn 2014; Asnake and Fana 2012; Tewolde and Fana 2014; Fana 2015; Kamski 2016a, b). These schemes and commercial plantations, although not realized on the grand scale originally planned,[13] are facts, and impact thoroughly on local conditions and landscapes (see Fig. 6.1). The assorted social, and human rights consequences are also significant,

Fig. 6.1 Lush traditional sorghum and maize fields of Suri near a village, 1992. A sight no longer seen in 2017. Note trees left standing across and near the fields

and probably will have a negative impact on long-term growth, prospects of 'inclusive' human development, sustainability, and durability. Environmental preservation and resilience are not guaranteed.[14] Suri informants have frequently complained about the fact of 'not being heard,' being seen as 'superfluous,' and having to conform to imposed livelihood and cultural changes so as to become 'modern.' They are painfully aware as well that their space is literally constrained: 'We have nowhere to go', they say, due to the presence not only of the state projects on their territory but also to gradually expanding neighboring groups that inhibit movement (e.g., of herds, or to new cultivation sites) and in-migration. The paradox is that the Suri have their own political–administrative district ('Surma *woreda*'), but this unit is powerless to act in the interests of the Suri community and cannot help them in maintaining their rights to land and decision-making. There is allegedly also endemic corruption, which draws in Suri members of the *woreda* council.

The new economic dynamics of Ethiopia is much lauded in global economic discourse and in donor country and World Bank development assistance circles, but is informed by unbalanced macroeconomic views, a neglect of the role of local economic systems in place (cf. Hallman and Olivera 2015 for a South American case study), and an absence of monitoring or evaluation by donor funders. This top-down and supply side-driven 'developmental state' approach is still little studied as it is unfolding 'on the ground,' at the local level; this is not really in the purview of donor countries and global institutions.[15]

Suri Livelihoods and Environment Use

The environment of most local people in the Ethiopian Southwest, including the Suri and Nyangatom, is dominated by transhumant agropastoralist livelihoods, but there is also river-bank cultivation, gold panning, and some hunting-gathering.

Among Dizi and Tishana Me'en people (northwest of the Omo), who live in the hills, there is a mixed agrarian economy, based on shifting cultivation, honey production, and gathering, with small-scale domestic livestock keeping (Me'en), or among the Dizi sedentary grain, *enset* and tuber cultivation, with occasional gold panning. Me'en have developed cash crops like coffee and sesame in recent years. Exchange and local market relations connect the groups, but with cattle raiding, theft, and clashes also occurring.

In the Suri areas (the lowland savannas and the foothills), we find a landscape with numerous cattle tracks, water points, ritual places, hunting-gathering domains and cattle camps, and more permanent village sites in the low hills. The tree- and species-rich savannah is (was) maintained due to Suri frequently moving cattle to prevent overgrazing, and their not totally uprooting vegetation and trees when preparing cultivation sites; they cut them only partially, to allow regeneration. In fact, the specific park-like landscape was created and maintained by Suri via periodic controlled grass burning, transhumant grazing, and the frequent moving of settlements and fields.

There was a coexistence with wildlife: Hunting was done since the early nineteenth century in the current Omo National Park area, but (until recently) not in predatory fashion. It was a system of hunting not imbued with cosmological or supernatural ideas about the 'harmony' of natural species, but more based on a pragmatic attitude, defined by long-term interest.[16] The local landscape is an integrated system of various types of land use and cultural use: not monotonous, but varied, interlocked, and filled with 'meanings and memories.' It is dotted with ritual and other culturally significant places[17] that make it 'home' (Many of them still unexplored among the Suri, and partially taken over and now inaccessible after occupation by the neighboring Nyangatom people).

The Suri area therefore being a typical agropastoralist livelihood zone, the people always relied on a smart *combination* of cultivation, transhumant livestock herding, and hunting-gathering. The lowland area (meaning in Ethiopia: below ca. 1000 m.) is overall food secure.[18] Cultivation is rain-fed (although at the Omo River a few Suri, imitating Mela or Mursi, also did flood-retreat cultivation (dependent on annual flooding bringing fertile silt deposits). Drought is rare, and varieties of sorghum used are attuned to the rain and soil characteristics of the fields (see Fig. 6.1). Maize also yielded relatively good harvests.

Most Suri households (among all three subgroups, Chai, Tirmaga, and Baale) have cattle in low-lying areas and are transhumant. Previous pasture areas in South Sudan were lost in the course of the twentieth century due to ethnic group conflict, enhanced after the 1980s by the spread of semi-automatic weapons among all groups, leading to an intensification of raiding and many hundreds of casualties over the past quarter century. Sharing of historical pastures areas of Nyangatom and Suri has steeply declined. Only in recent years, there is some contact again between the two groups (and with the Toposa in South Sudan) about asking permission to graze cattle in the respective border areas. Food insecurity when it occurs is due to the effects of livestock raiding—leading to serious decrease of milk (products) and cattle blood available for consumption as well as to wealth loss (having a cascade effect: less trade of cattle or goats for food, delay of marriage, and decline of payments).

For the long rainy season, starting in ca. February, Suri households plant sorghum, maize, and beans in the fields, and in the gardens around the house cabbage, spices, sweet potatoes, pumpkins, and some cassava. The gardens are the full responsibility of the women, but they also take a great role in the maintenance and weeding of the larger fields for staple crops. While the Suri economy is (was) largely self-sufficient in food due to a solid cultivation basis and products from livestock (meat, milk, blood; cf. Abbink 2017a), they also sell cattle and goats in local markets for cash, if need be. Their main cash income in the last 25 years, however, has been from the sale of alluvial gold, which they pan in the many streams in the area.

The main markets for the Suri zone are in the small towns of Jeba and Maji, as well as the frontier town of Dima (although insecurity there is high). Town traders also buy livestock and gold in a kind of 'contract' arrangement, i.e., often before it reaches the physical market. Suri buy additional food supplies, canisters, razor blades, soap, and alcoholic drinks (*araqé*).[19] They tend to avoid selling their cattle, in order to maintain their herd sizes. They usually only sell adult male animals (older bulls, oxen), never cows or heifers. Prices for all products can fluctuate significantly due to all kinds of factors, including insecurity. In May–June, just before the main harvest, there is a 'lean' period, when food is in short supply. At this time, households tend to purchase additional food with income from the sale of livestock, and also gather more wild food items occurs (seeds, nuts, and fruits). This 'gathering' component of the local economy is often seen as a sign of a 'primitive economy' by state officials and development agents, but the products are nutritious and plentiful and an integral part of the Suri diet.

Next to population growth, climatic variability and spatial competition as underlying causes, brief lapses of food insecurity were produced by group conflict and by occasional livestock diseases (e.g., rinderpest, pasteurolosis, blackleg, contagious bovine pleuropneumonia, and foot and mouth disease). Pasteurolosis occurs particularly in October–December and blackleg during the rainy season (cf. FEWS-Net report, ibid.). While Suri cattle overall are healthy, households could obtain drugs if needed either from occasional government veterinary service in Kibish (the main town of Surma district), via a Protestant-Evangelical mission organization located in Tulgit town, and sometimes on the black market, from pastoralist traders in South Sudan.

The Suri traditionally diversified their productive activities, geared to environmental possibilities and to mobility of livestock as well as cultivation sites (changed after a certain number of years), and they make low-intensity but optimal use of the natural conditions, with simple technology. Mobility of herds, following the best available feed sources, is a *production strategy*, not a 'coping strategy' to deal with 'problems' (cf. Krätli et al. 2013, 44).

Ecology, Space, and Place

It can be noted that the cultural ecology of the Suri in particular and related peoples reveals an adaptive system of agropastoralism geared to the three pillars of livestock rearing, field rotation agriculture, and hunting-gathering. Gold sales and petty trade are activities that have become an essential addition to their economy. But since a decade or so the gold trade, in which they were the pioneers in their own area, is threatened by the influx of all kinds of non-Suri newcomers, mostly highland Ethiopians, who aggressively compete and push out the Suri from their traditional places, with the help of armed forces and police.[20] This appears to be another phase in the gradual disempowerment of the Suri: next to the pressure to reduce their herds (and thereby their capital) and to give up their agricultural fields for small plots near newly villagized locations, also their chief cash source (gold) is being taken from them. Since 2012, dozens more killings and cases of robbery of Suri gold miners have occurred. In addition, the way the highlanders do gold mining is much more damaging to the environment, as they use mercury (and sometimes arsenic) to 'purify' and separate the gold from rock and stone. They also dig deep holes all over the area, in contrast to the Suri, who do mostly surface mining and leave the landscape more intact. During 2017, it also seemed that foreign investors would be allowed to mechanically mine Suri gold places in the south, undoubtedly set to contribute to a further deterioration of the environment, and taking away resources from the Suri economy.

The Suri's evolved local livelihoods were imbued with cultural knowledge that represented experience-based strategies to survive, diversify risks, and adapt to changes. These are threatened now due to the livelihood transformations imposed upon them. With neighboring peoples, there is a dimension of competition and conflict which has become more serious since the 1980s due to one or two extreme droughts (e.g., in 1984–1985), group conflict (impact of the South Sudan civil war since the early 1980s), emerging climate variability, population growth on all fronts, and growing state interference. The latter has not diminished the conflict potential, probably the contrary.

Definitions of culturally and economically meaningful 'space' and 'place' as perceived by ethnic groups are vital in this area, as they reflect access claims to land, water, and pasture, symbolic constitution of 'homeland,' and economic range (e.g., extent of territory or pastures). The elements have a direct environmental dimension. The ethnology of 'space' and 'place' has been developing since ca. two decades at least (see the book by Hirsch and O'Hanlon 1995; Low and Lawrence-Zúñiga 2003), and has predominantly focused on the changing cultural meanings and interpretations of these two concepts as locally embedded and articulated. In the case of the Suri and other peoples in the Ethiopian Southwest, it is striking to see how the state politics entering the region has undermined local conceptions and practices regarding space and place, subverting local cultural narratives, and social cohesion.

Under the post-1991 ethno-federal state, the Suri were assigned their own administrative unit: the 'Surma *woreda*' within the Bench-Maji Zone. That is, their 'identity' is officially recognized and laid down via territorial anchoring. Without such an anchoring, no group in today's Ethiopia has a chance to be recognized[21] or—in the long run—to survive. So there is a specific 'Suri territory,' although in the Suri mind, their territory is wider than that: extending to all areas where cattle can find grazing and water, and historically much beyond of what it is now. The obvious problem is that 'Suri territory' is now contested by the federal state, which wants 'the resources.' It 'nationalizes' the environment and the land, which is constitutionally defined as state property. The idea of all land as 'public,' i.e. state owned, has led to the notion among state planners and administrators that the environment and its 'resources' do not have to be factored in as 'costs': They are freely available. It is a familiar story and applies of course also to other parts of the Omo Valley, with the Nyangatom (south of the Suri), the Mursi, the Mela, and Chirim (Me'en) people all in danger of being pushed out or challenged by expanding sugar plantations and commercial farms (cf. Avery 2012: 59, 2014; Fong 2015; Kamski 2016a; Wagstaff 2015). The Surma *woreda* authorities (with some Suri officials, but mostly outsiders placed there by the ruling party) have no say over what is happening in their district and cannot prevent or modify plans imposed by the Regional or the Federal Government. The latter's approach thus also subverts the political formula of 'ethnic federalism.'

For Suri, the environment is a *cultural* landscape, the product of interactive engagement over centuries. Land and nature are seen as an open space or 'resource' to be shared for all, but people need to respect and 'maintain' it. There is land classification and use planning, and no idea of 'free riding' of humans on nature. Suri, like the Mursi (Olisarali and LaTosky 2015), make land use decisions for the collective, decided and confirmed in public debate assemblies of adult males as well as in more detail within the five or six cooperative herding units (*buran*), that all have a certain territorial range. Their classification of land use is into roughly four kinds: (a) lowland plains, space for livestock grazing and moving/exploring ('open space,' no real limits), with cattle camps, and basically forbidden for married women, (b) cultivation areas, nearer to villages, and horticulture plots (women's domain) in the villages, (c) areas of bush and forest for gathering and hunting, and (d) ritual and public spaces, such as burial sites for ritual leaders (*komoru*); initiation sites for age sets and a new *komoru*; public debate sites). Suri used to plan and decide on their land use in a long-term, 'interactive' perspective—where to make new fields for what, where to go for herding, which forest/bush area to leave alone, etc.

Although these local societies like Suri, Nyangatom, Me'en, or Dizi were not necessarily 'well-integrated societies' marked by ecologically responsible interaction with the environment, there was no tendency to exhaustive over-exploitation of the 'natural resources,' also because the technological means and the economic inequality structures to allow or fuel this were not there. Over-exploitation usually started (in recent decades) due to external stimuli, population growth, and imposed territorial restrictions (Fig. 6.2).

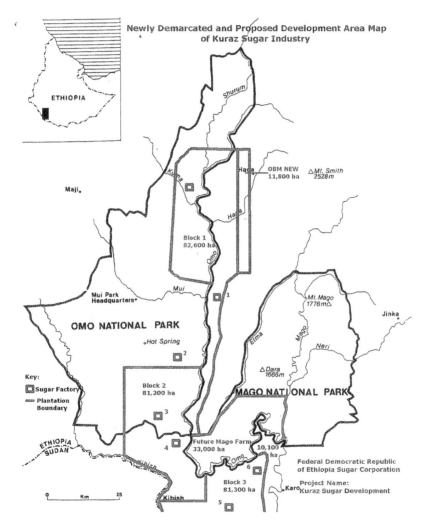

Fig. 6.2 Plan of the sugar plantation surfaces in the Lower Omo basin (*Source* https://www.survivalinternational.nl/nieuws/7865, from an official Ethiopian project document, 2011)

State Development Views and Plans

Today, critical issues of sustainability, access, profit, and future use of the land and its resources are predominant among all groups, as space is getting limited, populations increase (notably among Me'en), and the new state economic projects now compete with the locals and their resource use. Local land-use access and 'planning' thus get increasingly difficult. The state interventions are imposed on top of the local subsistence economies and are non-negotiated, interfering with cultural–ecological practices in place. Residents, notably Suri, compare the indifferent and often abusive policy of the state (cf. HRW 2012; Oakland Institute 2013) toward local inhabitants with its aggressive intervention in the local physical environment—damming the river and holding up the water, slashing all vegetation, roads crossing the area and the fields, construction of moats around plantations—they see parallels.

The Omo Valley—like a few other 'frontier regions,' like Gambella (cf. Seyoum 2015 or Benishangul-Gumuz, cf. Tsegaye 2016)—is thus an area in flux: infrastructure building, environmental overhaul, agrarian industrialization, and population movements. For instance, in the Kuraz sugar plantation plans (Kamski 2016a, b), the settlement of an imported labor force of ca. 400,000 people from other parts of the country was foreseen (among them, resettled Konso). This would lead to new 'urban centers,' without much connection to the hinterland, and friction with local groups. While these initial figures have been much lowered in recent years, the tensions are already there. Domination and displacement of local people, who are seen as having no economies of scale, no food security and no 'skills', are proceeding.

The state political and economic expansion started in full force after the Ethiopian leadership (the late PM Meles Zenawi 2011) decided to tackle the perceived 'developmental deficit' of rural Ethiopia in a radical way. The plans did not lack vision and ambition—probably there was too much of it. Part of the first (2010) 'Growth and Transformation Plan'[22] was a major boost of the energy production infrastructure and agrarian export production via investments in hydro-dams, roads, and large-scale mono-crop plantations on leased land in 'empty' or 'under-used' areas. Meles's vision of the agropastoralist peoples was outlined in his controversial speech of 2011 (Meles 2011).

The state presence—government administration—in the towns is established, in addition to the investment projects mentioned above, and the extraction of resources is on the rise (coffee and sesame, minerals, wood, and gold). Local people struggle against the effects of this alternative land use and livelihood competition, but are met with refusal and often arrest. No reasonable, open debate, let alone criticism of the plans, is tolerated.

One stated aim of the massive agrarian investment schemes started since 2010 was to 'bring development and civilization' to the Southern peoples. It was apparently envisaged (cp. PM Meles's speech of 2011, cited above) that they would be laborers, technicians, etc. on the plantations, and that 'technology transfer' and the like would benefit local agriculture. This has not happened. At the most (so far, in late 2017), some members of the local ethnic groups (Suri, Nyangatom, Mursi, Mela) were employed as guards or drivers, e.g., at the Salamago plantation east across the Omo. Their salaries were rather high according to local standards, but the work is often temporary and no one was trained for higher functions, not even the few locals who received a BA degree in accounting or in other economic–administrative skills, and very few gained a position of influence. On the plantations, the tensions between members of different ethnic groups are high. There is much robbery and killing of people for money, and people also disappear.

A key issue is that the state interventions do *not* have the aim to strengthen local food security, but primarily to generate 'export value' (energy via the Gibe-3 dam, cash for the state via the sale of sugar, cotton, biofuels, etc., from irrigated fields). In fact, local systems of food production are undermined.[23]

A closer look at the Ethiopian state's view shows that, as many other 'developmental states,' it projects a view of the environment or the natural conditions, as an 'enemy' to be conquered, in a classic nineteenth-century Western sense. It has to be 'overcome,' to be dominated and exploited. The rhetoric of Ethiopia's GTPs I and II is full of terms reflecting this, and the frame of thinking is handed down along the various administrative levels, whose officials have to conform to it and are judged on the 'targets' that are set.

State agents often show a lack of valuation of ecological conditions as 'production factors' with certain costs. That is, there is serious 'underpricing,' most clearly seen in the case of the construction of the massive sugar plantations. It fits in this approach that local people's identity and territory are to be 'decoupled' in the development plans and even by legal means: All land is state property and no intrinsic or symbolic bonds are recognized by the government between peoples and their eco-habitats. Neither have proper geo-ecological feasibility studies or impact assessments been made; and if there are, they are not accessible. Local people do not really have a voice, and there is no educated elite to plead their cause.

The Suri, in so far as they have explicitly conceptualized their relation to the environment, connect to their natural physical conditions in a more accommodative, pragmatic, and adaptive way, that takes a longer-term view of mutual dependency: They know they have to maintain it. But, as noted above, this does not automatically mean that they are 'in balance with the environment' or see themselves as 'custodians' of nature. They have overgrazing in places and sometimes excessively hunt in certain areas. Still, they do not share the state views (in the dam and plantation areas) that land and nature are like a 'free resource' to be exploited without limits.

Specifically, the state's land use ventures are far-reaching and have significant environmental effects, the consequences of which are not yet entirely clear. First and foremost to consider is the impact of sugarcane plantations and their expanding irrigation demands. Ca. 175,000 ha. of sugarcane plantation was foreseen in the 2010 plans (Avery 2012, vol. 1, 12f., 52).[24] As of late 2017, huge plantations are in place on the east bank of the Omo, on land of the Kwegu, and the Mela and Chirim (both Me'en groups), eventually to cover 82,600 ha.,[25] expanding south to the Mursi area. These people lost virtually all their land and resources access, and many of their cattle (due to lack of pasture or access to the Omo River). The riparian forest was destroyed over dozens of kilometers along the river, and with it, natural species, wildlife, food and ethno-medicinal plants, and places to hang beehives or as dry season pasture and shade for livestock. River-bank cultivation—a key food security element—was lost.

The effects already reached the Omo delta as well. The Nyangatom and Dassanech reportedly suffer from the cessation of the annual floods and the drying out of the area due to closure of the Gibe-3 dam and the receding water levels of Lake Turkana (cf. Carr 2017 for an analysis). As of late 2017, the lake had already dropped 1.5 m (Source: HRW 2017). In general, the rapid decline of forests, added to that in other areas of Ethiopia, contributes to greenhouse gas emissions, and this is not compensated by the sugarcane field greenery.[26]

Since 2016, sugarcane plantation construction has also started on the west bank of the river, down from the Sai mountain, right in lowland Suri territory, near Udumt village (Suri). It will reach down to the Dirga hills area, which will be just outside the plantation area.[27] In these Sai plains ca. 2000 Chai Suri live with their herds, grazing in the plains below Sai, and others from the west also come to graze cattle. Their pastures and their access to the Omo are now being disturbed. The Mela scenario of environmental decline and social upheaval (see Buffavand 2016, 2017) is likely to be repeated, as livelihoods and food security are threatened.

What are the environmental effects of sugarcane plantations?[28] As in the Mela area, in the Suri and Nyangatom areas west of the Omo—eventually to cover a staggering 81,300 ha.—their construction is leading to natural habitat clearance, with few local species of trees and bushes retained. A dramatic reduction of biodiversity and of food plant availability for gathering and ethno-medicinal purposes likely occurs. This is even apart from the loss of fields and pastures. The associated irrigation of the sugarcane fields requires lots of water, also taken from the Omo river. This will likely even lead to an overuse of water and to salinization. There is also a danger of soil erosion due to the intensive, yearly cultivation of one crop. Especially in this area, soil fertility is uneven (as in many parts of Africa, see IFAD-Montpellier Panel 2014) and prone to decline. Substantial irrigation and fertilizer addition will be needed year in, year out, and the risk is soil quality loss with long-term effects, especially in the absence of other vegetation. Also, the discharge of sugar mill effluents will be negative and is already making its effects felt in the Block I plantation areas in Mela territory (Salamago *woreda*). In principle, the residue of harvested cane could be collected and fed to livestock of the neighboring pastoralists, but this is not done.[29] Instead, the sugar plantation authorities decided to use it for additional energy generation. Neither the wood from the thousands of trees cut on the plantation sites could be used by local people. The negative effects of intensive use of chemicals, pesticides, and runoff pollution that have been reported of most large-scale plantations can also be expected in the sugarcane area. They have already been noted on the east bank of the Omo (Kamski 2016b, 6), e.g., among the Mela people.

The restrictions on the use of space—less transhumance areas for herds—leads to pastoralists being forced into smaller areas, where overgrazing takes place. Some of the Suri near Sai mountain west of the Omo have for this reason even moved across the Omo. Furthermore, despite promises that water would be provided to the new agrarian plots near 'resettlement villages,' according to reports from local informants (2016–2017), the irrigation channels from the (lowered) Omo river are withholding water from grazing areas and fields.

There is some irony in that the massive investments and clearing of natural forest and habitat and the projected mass settlement of import-Ethiopians from densely settled north-central areas will lead to rapid environmental decline and erosion, features that are now recognized in northern Ethiopia and are eliciting counter-measures (cf. Abbink 2017b). It seems that in the 'empty' Ethiopian Southwest first such development-induced erosion and soil degradation processes will be allowed to occur, before any 'mitigation measures' will be considered.[30]

Other agrarian investments also proceed, next to the state sugar plantations, and will become more numerous. In the heartland of the Suri area west of the Omo, there was already a foreboding of things to come with the construction of a *31,000 ha.* plantation in Koka locality, northwest of the town of Maji after 2010.[31] Although it was only operational from 2010 to 2014, run by a Malaysian company that had leased the land, it led to massive 'clearing,' huge enclosure, and cutting off access to pasture or to the waters of the Koka river for Suri herders. The Koka plantation was not successful due to a host of problems (cf. Wagstaff 2015, 19–20; Oakland Institute 2014a): insufficient preparatory field studies were done by the company, the crop choice (sesame, palm oil, mushrooms, and rubber) was over-ambitious, there was lack of capacity and resources, operational costs were very high, and there were some security problems. Local Suri herders and cultivators were not consulted and did not accept the confiscation of their dry-season grazing land and transhumance routes. There were repeated skirmishes with the guards, whereby people were killed. Also, the tensions between Dizi people (supporting the plantation) and Suri increased, allegedly due to issues of tax collection from the Malaysian company: Dizi demanded any company tax be paid to Dizi *woreda*s (Bero or Maji), and Suri were demanding that they should also profit.[32]

The Koka plantation was partly on Dizi and on Surma *woreda* lands. The plantation was guarded by Ethiopian security personnel, and a huge moat was constructed in the fashion of a medieval castle to keep Suri and others out. Cattle would fall into the ditches, leading to anger among Suri owners. Frequently Suri people invaded the plantation, e.g., to take maize. While the plantation was abandoned in late 2014, the land was partly retaken by Suri, building several cattle camps there. But it was then scheduled to be given to other investors, again without consultation with the local Suri resource users, and additional armed units arrived to protect the area. In 2017, a number of Ethiopian investors from outside the area were promised the land, and they started new plantations.

A significant number of violent incidents occurred in the last decade, partly due to the plantations generating resentment and being 'militarized,' and also due to the ongoing, ill-managed ethnic group tensions on the presumed 'benefits' and the controversies on land use and access. The growing insecurity and chaos in local economic relations also produced more banditry in the area. First, the Suri were seen as the exclusive perpetrators, ambushing Dizi, and Me'en and villagers. They were often involved, but gradually it appeared that more and more Dizi and highlanders were also active, often masquerading as 'Suri' and even going so far as to dress like Suri and apply white paint to their faces. The local authorities of the Bench-Maji Zone (the unit where the Dizi and Surma *woreda*s are located) have had some success in recent years to curb this violence by organizing peace meetings and amnesty sessions. At the same time, the authorities try to disarm all local people, especially the Suri.

In the background of the half-hearted state efforts and promises on 'development' to local peoples is the wish to ultimately *eliminate* transhumant pastoralism and have all the people settle—the familiar old cultural scheme based on mistaken perceptions, if not disdain, of smallholder farming and agropastoral economic strategies, this despite the 'constitutionally guaranteed' right of pastoralist peoples to their livelihoods.[33] Suri informants (2015–2017) mentioned the repeated exhortations of local administrators—often issued with threats—that they had to (re)settle, reduce herds and cultivate maize (as well as 'give up bad customs'). Ideas from the Suri themselves routinely neglected, and those who protested were arrested. There is even no thinking among state officials about rangeland development for cattle-herding.

Effects of State 'Landscape Reconstruction' on Local Peoples: More on the Suri Example

Suri livestock herding, their core activity, is relegated to a smaller territory, and there is pressure to reduce animal numbers. They are forced to limit transhumance, leading to mounting resource pressure. Data from c. 22 Suri households that were followed for over more than a decade show that they not only all lost animals to raiding by Nyangatom and Toposa but also faced greater problems in finding reliable pasture and water for the animals in general. Counter-raiding did not compensate for the losses, and it incurs police/army action. Security forces rarely if at all assist the Suri in self-defense or in recuperating cattle raided by Toposa. The pressure on Suri to congregate in the state-designated 'resettlement villages'—a movement not devoid of coercion (cf. Wagstaff 2015)—has the consequence that herd mobility is reduced, as cattle cannot be kept near the villages. They continue to herd them in the plains, but the space is shrinking; they no more have 'buffer zones.' There is movement toward border areas near Nyangatom and Toposa, but with permanent risk. The movement up north (to the Akobo River valley), necessary despite the lesser quality of the pasture, was thwarted by the closure of the area covered by the huge Koka plantation (see above). So what are the main effects?

- Restrictions on livestock herding and losses incurred by raiding mean impoverishment. The raiding in the area has not diminished in the wake of the development schemes but continues, and animals are rarely recovered by local police or armed forces. There are tragic stories of people brought to poverty in one day. In 2016, one of the Suri *komoru*s (ritual leader) said in a soft, low voice: 'All our animals were taken, all. We have been made poor. Nothing is left.'
- Suri seriously resent the 'settlement programme' (Amharic: *säfära program*) of the government obliging them to sedentarize and congregate in limited spaces and finding neither room for the livestock herds, nor for proper cultivation plots. The type of housing—tin-roofed square structures of bad quality—are impopular, and people do not understand why the new, prescribed village sites are not built near a water source, like traditional Suri villages are.[34] They also resent the patronizing efforts to 'teach them how to do agriculture,' while they are excellent cultivators already, both of food staples (sorghum, maize) and horticultural crops. They also see dangers of over-exploitation of the limited territory. There are complaints about the density, the cutting down of shade trees and the rapid exhaustion of the small fields for cultivation. Decisions as to plant what and where by Suri are subverted, leading to confusion and restriction of production. This shows that the 'villagization programme' has not been thought over properly regarding its impact on the immediate environment. In addition to the villagization and plantation outlays, the area is crisscrossed by road-building projects, both done for security reasons (to allow army access) as well as facilitate external investors moving in. Suri fields and pasture routes are crossed by these roads, which tear up the landscape and are resented.
- Suri persistently comment on the decline of variety and numbers of natural species (vegetation, wildlife, food plants and medicinal resources, and even fish in the Kibish and Omo rivers), especially in areas close to the sugar plantations and other large-scale agrarian projects. They note that wild plants they use for one of their main dishes are decreasing and also that wildlife is 'pulling out' or disappearing.[35] At the same time, they abandon restrictions on hunting; when outsiders show a demand for ivory and other wildlife products like

leopard skins, they increase hunting to sell to them.[36] This is done in response to the decline in other livelihood domains and changes the Suri attitude toward the environment.
- There are also observations made by local people of the increase of invasive species like *Parthenium hysterophorus*, *Prosopis juliflora*, and others (cf. Berhanu and Nejib 2016 on the current situation), which have done such enormous damage in the Afar pastoralist area in northeast Ethiopia (cf. Lisanework et al. 2010; Rettberg and Müller-Mahn 2012). These disastrous weeds are already invading Nyangatom areas and will likely spread in the wake of the construction of the sugar plantations. In the literature, it has been noted that the heavy construction equipment (bulldozers) is often spreading the seeds of such plants.
- There is also more movement of dispossessed Suri people who lost land and pasture or cattle and search for places of safety. Some have lost large numbers of cattle due to raiding and moved to new areas to try and recover via cultivation and goat herding in border zones, and to avoid the obligatory 'settlement areas.' But resource pressure is augmenting, because space is literally shrinking. In the past 5–7 years, Suri moved closer to the southern end of Dizi territory, the Kolu hills, due to being pressurized by Nyangatom and Toposa. Suri settlements remain unstable due to the security threats. Displacement and land loss is a general problem in the Lower Omo Valley project area and has also affected the Mela, Kwegu, Kara, and Mursi people east of the Omo, as well as the northern Me'en, e.g., near the village of Biftu, west of Mizan town.[37]
- Alternative futures are being planned: Local people including Suri are to be employed on plantations as wage laborers, and to be given 0.25–0.5 ha. plots of land for subsistence cultivation near their designated 'villagization' area. Suri—like the Mela across the Omo (cf. Buffavand 2017)—find this preposterous and reject it; it interferes with their economic decision-making, and they say that the area around such villages has too little cropland and is soon depleted of wood and other resources (see above). The experience so far with the few existing new villages bears this out. Some Suri state that it is unjust to be made 'unemployed'—by denying them their herding

lifestyle and cattle wealth and trying to limit their cultivation practices—so as to be forced to be laborers on a plantation. They think also that the restriction of mobility and means is sub-optimal use of the environment and not conducive to well-being. Several Suri have indeed started to work on the plantations, e.g., the one in Salamago, east of the Omo. This was often due to their having lost all their cattle in raids. The plantation work, however, is dangerous and is done in bad circumstances; the tensions between members of different ethnic groups are high, and there are much robbery and killing of people for money, and people also disappear.
- A survey of Suri households[38] in Surma *woreda* indicates that overall food security, especially in areas of partial villagization, has declined, as has crop and diet variety. This may be due to the prohibition to cultivate freely and the pressure to decrease livestock herds. The 'free supply' of maize, as a premium to get people to settle, did not prevent this.
- A general feature of the developmental offensive in the Lower Omo Valley is the belittling of local peoples: They are to follow orders and have no say in implementation or in offering alternatives adapted to local conditions. Many sources have mentioned gross abuse of locals who protest or even dare to ask questions on what is happening. Local young spokesmen or emerging leaders were frequently arrested and killed.[39] This issue, however, needs more research.

The state sees the southwest as an anomalous political space, still without proper authority or leadership, and as a domain or territory yet outside 'the law.' State actions and territorial rooting via administration, police posts, and land schemes *establish* its presence and produce locality for the state, so to speak. In the process, which shows obvious parallels with historical cases of state-making elsewhere, local people are subjugated or 'disciplined' in a new order—but so far this did not give them active citizenship rights (cf. also ILO 1991). As noted above, the state authority claims to space and environmental 'resources' has led to relegation of locals to second-class status and to exclusivist practices that break the existing practices of environmental management.

The political ecology of development and environmental exploitation as initiated by the Ethiopian state since ca. 2010 has thus clearly changed the ethno-landscapes and (implicit) ecological management rules of local peoples in the Omo Valley. These eco-scapes are being fragmented and divided into unconnected parts and reduced in viability and diversity. The water and vegetation systems of the Omo Basin are disrupted, and this impacts on overall fauna and flora. Consequences across the board have not been favorable. While population pressure and climate change are also marching on in the Lower Omo Valley, they do not in and by themselves produce environmental problems and local food insecurity. The large-scale agrarian plantation economy set up in the area, coupled with (re)settlement policy, leasing out land to numerous outsider-investors, etc., is contributing to instability, as it is not sufficiently geared to environmental conditions and political consensus. Sustainability and local acceptance would demand contextual planning, better choice of suitable crops, densities, and building on proven local economies, like already existing staple crop production and river-bank flood-retreat cultivation (which was much more productive than the current irrigated agriculture on plots near the new villages).

Conclusions

Suri are among several peoples in Southwest Ethiopia facing environmental problems as a result of state landscape reconstruction, forced livelihood changes and resettlement, and partly due to longer-term climatic changes. State misconceptions on massive-scale agrarian development, free riding on nature, and non-negotiated resource extraction enhance the problems, undermining the ecological resilience of the Lower Omo River basin. The changes are quite far-reaching and can best be studied via their impact on local peoples that have evolved and adapted in the area for hundreds of years. Many of them, notably the Kwegu,[40] Nyangatom (Carr 2017, 145f), Mela (Buffavand 2016; Stevenson and Buffavand 2018); and Suri, are 'losing the plot' in a double sense: Literally, their pastures and cultivation plots and the freedom to clear them and exploit them where they want, and the sociocultural plot, due to mounting fragmentation and disarray in their societies—for the Suri, partly internally (Abbink 1998) and partly externally generated (Abbink 2009a, b, 2017a).

The state developmental ventures in the Lower Omo Valley—still informed by a model of post-Communist command-economy governance—have thus not significantly contributed to well-being and development (in its multidimensional meaning) but instead to more environmental assault and to more human misery. The number of people displaced, killed, injured, and traumatized in the past two decades is remarkable; in fact unprecedented. Suri lives are not positively impacted in either a social or a material way and show no good answers to cope with the problems. Only a few younger people that are inserted into the national education system and have found jobs and other benefits outside the agropastoral economy see advantages: A minority that is somewhat estranged from its own community and also is reputed to have gotten itself deeply involved in corruption (notably in the Surma *woreda*). They also keep a foothold in the pastoral economy, investing in animals. Overall, the Suri agropastoral way of life is thus under siege: Their lands—as those of the neighboring 'indigenous' groups in the Omo Valley Basin—are wanted by outsiders and their livelihoods are threatened. A convenient political and sociocultural labelling of these local groups as 'backward' and 'having nothing to contribute to the national economy' is exercised, buttressing the state's appropriation of lands and economies.

The process of change breaks the bonds of sustainable resource use that were predominantly in place among local ethnic groups (despite the occasional crises). The neglect or misunderstanding of local views and cultural values attached to space and place as enacted by populations in the Omo Valley and that define 'environment,' livelihood 'resources' and 'home,' tends to exclude them from (a) productive involvement in developmental ventures, (b) recognition as capable cultivators, and (c) citizenship rights; and finally, is detrimental to the local ecological system: Because the environment is seen as an 'adversary' to control, not a sphere to manage and live in. Suri had meaningful cultural templates to guide their interaction with the environment that helped them adapt and survive so far. These are denied relevance by state agents and are under pressure, e.g., in state schools, religious change (missionary groups) and political indoctrination. Still, it is likely that these cultural factors remain of influence and will resurface. As M. Sahlins has said, 'culture' is not a disappearing object (2000), and the Suri, who remain remarkably attached to their land, will likely try to 'encompass' the emerging new state order 'within their own cosmos' (cp. Sahlins 2000, 202).

The socioeconomic processes in the Lower Omo Valley so far, engendered by the Ethiopian state and its agents, including the investors and sugar company ventures, have shown severe political and ecological problems, reflecting those elsewhere in Ethiopia and the Horn. The environmental issues (including those directly related to the Gibe-3 dam, not treated in-depth here) are underestimated and are affecting local economic and societal resilience (cf. Turton 2012). Nothing was learned by the state from experiences like the Afar country debacle in the 1970s–1980s (cp. Behnke and Kerven 2013; Carr 2017, 27). The neglect of local adaptive systems in place, notably agropastoralism, and the lack of space accorded to them to *operate*, is dramatic, and enhances vulnerability on a wider scale. Next to the declining environmental conditions, the constant conflicts in the Ethiopian Southwest—with many people killed and injured—and the negligible growth of local productivity, incomes and well-being so far, may show that government policies are worrying and economically not working well. They do not deliver but make things worse and are very costly.[41]

There is a need for a context-sensitive *agro-ecosystems approach* to rural development and food security provision, as well as a new chapter in accountability and in 'stakeholder' involvement in more democratic and gradual processes of developing the local economies on the basis *of what is already there*. The so-called traditional knowledge—or rather, the 'experiential local knowledge' of people that lived in the place for centuries—has adaptive value and needs to re-evaluated and *integrated* into the agrarian modernization drive (Giles 2005b). This ecosystems knowledge can even be recognized in its financial value, so far just bypassed by governments (cf. Giles 2005a). Indigenous knowledge is certainly not all scientifically acceptable, but it neither is backward or useless (cf. Hendry 2014; Amborn 2012; Pardo and Macía 2015). Livelihood complexes in place (cf. Getachew 2014 on the Gedeo in South-central Ethiopia) can be seen as *biocultural heritage traditions*, rooted in specific ecologies, and functional (IIED 2014). They are multidimensional and deserve recognition and maintenance, even if they are subject to change and are partly adaptive to it. These traditions can be used as a basis for expanding the local economy and for incremental modernization. The heritage of the peoples and landscapes in Southwest Ethiopia can also be seen in this light. Livestock raising in these semi-arid lower areas of the Omo Valley, especially when more invested in, is a better option than the costly irrigated agriculture complexes that use massive amounts of water from a dwindling Omo River and are taking a heavy toll on the local soils, landscape, and biodiversity. The latter option makes tens of thousands of people dependent on mono-crops like sugar, which are subject to serious price global fluctuations, impoverished the ecology, thwarted local (river-bank) agriculture, and inhibited agropastoralism. Available environmentally rooted livelihoods and cultural complexes are too important to discard in the context of ill-prepared developmental schemes that are uncritically derived from 'Western' or global economic models and frameworks. Systematically addressing and incorporating these living heritage traditions, e.g., via a new investment approach, would ease the transition to new economic pathways and to new forms of place-making. In view of the big state interests, the political ecology of resource control, and the limits of governance in southern Ethiopia today (struggling with systemic corruption), it is unlikely to come off the ground, but it would be one trajectory toward countering the emerging environmental crunch.

Notes

1. The most critical and detailed indictment of the effects of the state-led transformation of the Omo River Basin comes from geographer Claudia Carr (2012, 2017). She in particular described the environmental and livelihood havoc wrought upon the Dassanech people in the Omo delta.
2. Cf. Smallteacher (2013), and Vidal (2015).
3. See already: 'Aid agencies turn blind eye to "catastrophe" in Ethiopia,' April 15, 2013 (http://www.survivalinternational.org/news/9125).
4. The Suri people—like many other ethnic groups in the South—have over the past two decades come to see their land or place (home territory) as a small dot in a wider world outside—due to state encroachment and cultural–ideological 'reform,' constant attacks from South Sudanese people like the Toposa, presence of tourism and missionaries, and development projects (cf. Turton 2005 on the Mursi).
5. This paper is based on field studies and post-field contacts with local conversation partners over the years 1999–2016, and information was drawn from observations, selected household surveys, interviews, and group discussions.
6. The Lower Omo Valley is also a UNESCO world heritage site. UNESCO has repeatedly called for protection of this area by the Ethiopian government (latest: UNESCO 2015).
7. Information from Abbink (2017a).
8. Maximum, it is usually lower: about 1100 mm, but about enough for shifting cultivation, especially of sorghum, in agropastoralist systems (Mursi, Kara, Kwegu, Nyangatom, Me'en, and Suri).
9. There are three subgroups: Chai, Tirmaga, and Baale. The Baale (self-name) speak a somewhat different language and are sometimes seen as a separate group. Part of them live in South Sudan (near the Boma Plateau). The three are therefore not really a unified ethnic unit. The exact number is not known. The 2007 Ethiopian national census figures are estimates based on an extrapolated sample.
10. See Buffavand (2017, 36).
11. Notably the local groups of Bayti, Boqol and K'asha.
12. Cf. the works of Silvester (2009), Feron (2008), or Temperley (2015). Cf. Abbink (2009b).

13. 'Sugar' may also be a risky choice for the future, seeing the fluctuating market prices, its more and more unfavorable image, and the emerging alternatives: cf. Cox (2014) (www.theguardian.com/science/blog/2017/nov/21/sugar-industry-withheld-research-effects-of-sucrose-50-years-ago-study-claims, accessed December 10, 2017).
14. Scientific research has made these points repeatedly, but with little impact in the policy sphere: Cardinale et al. (2012), Barbier (2014), Gross (2016), Masood (2015), Newbold (2015), Pardo and Macía (2015), etc.
15. See also Oakland Institute's report of (2014b).
16. The cosmological connections as recognized by the Mela people, across the Omo (cf. Buffavand 2017, 288–291), were not in evidence among Suri.
17. Well-explored for the related Mursi people in work by Brittain and Clack (2012), Brittain et al. (2013), Clack and Brittain (2011a, b, 2017) and for the Mela by Buffavand (2017), 215–234.
18. As the report on the 'Surma Agro-Pastoral Livelihood Zone' in the 2006 USAID *FEWS-Net report* (section 'SNNPR Livelihood Profile') also states (p. 1) (www.heawebsite.org/countries/ethiopia/reports/hea-lz-profile-surma-agro-pastoral-livelihood-zone-sdp-snnpr-ethiopia, accessed June 5, 2010). But, there were a few exceptional crisis periods: see Abbink (2017a, 124).
19. Cellphones and accessories also became popular.
20. Cf. Wagstaff (2015), 25 on two mass killings on Suri in gold mining areas in 2012.
21. Cp. the very telling and in fact tragic case of the Gabbra, described in Fekadu (2014).
22. http://et.one.un.org/content/dam/unct/ethiopia/docs/GTP%20English%20Vol1%20(1).pdf (accessed February 20, 2017).
23. An extra underestimated effect of hydro-dams (like Gibe-3 and possible future dams) may be their contribution to short-term global warming: see M. H. Hurtado 2016, 'Dams raise global warming gas,' (www.scidev.net/global/energy/news/dams-raise-global-warming-gas.html, accessed November 10, 2017).
24. And up to 375,000 ha. if all future state and private plantations are added up.
25. More than half of it done by late 2017.
26. Ricketts et al. (2010, 1) estimated already that the destruction of forests worldwide accounted for ca. 15% of the greenhouse gas emissions.

While Ethiopia formally acceded to the emission reduction conventions and to donor country plans on maintaining forest cover and biodiversity (and has contributed to this in the past), it has now chosen for a developmental path that pushes virtually all environmental aims and costs aside (cf. UNESCO 2015, 17). Thus, in practice, no environmental recovery or biodiversity policy is seen on the ground (despite the nice plans: FDRE 2005, and despite advice from its own Wildlife and Conservation Authority: EWCA 2010). A new start was made—also donor-funded—with the 'IGAD Regional Biodiversity Policy' plan of 2016 (IGAD 2016), but again there is no evidence of meaningful consultation or understanding of local peoples as 'stakeholders,' and the external conservation vision on landscape and environment predominates. On p. 7 of this document it says: 'Member States shall promote joint management of transboundary and shared biodiversity resources and Protected Areas, involving local communities at all times.' But so far, little evidence of this is seen. There is no discussion either on how the ambitious plans for such biodiversity maintenance will articulate or clash with the 'development' plans of the IGAD state governments.

27. In the Southern Block, the Nyangatom area and part of the Omo Park will be covered, right up to the southern part of Dirga.
28. An excellent study on which I rely here is WWF (2004).
29. As evident from a recent internal report (see FDRE Policy Research Centre 2017), the management of projects like the Omo Kuraz sugar project is also riddled with corruption and mismanagement of public funds, which takes its toll on human resource management, rights of local people, and on the environment. Cf. also *Addis Standard* (2017).
30. In addition, such resettlement from one region to another also skirts around the national need for a strong policy on population growth control.
31. First mention: Mehret Tesfaye, 'Ethiopia: Malaysian investor launches 3.7 b. *birr* palm oil tree plantation,' *ENA News*, May 9, 2009 (www.ethiopianreview.com/articles/4257, accessed July 5, 2011).
32. This was based on misperceptions and distrust. There *was* no tax to be paid because the government had given the company a tax reprieve for the first five years.
33. And laid down in the FDRE's *Rural Land Administration and Use Proclamation* (Procl. No. 456/2005), *Negarit Gazeta* (Addis Ababa) 11(44), 2005.

34. A promised pipe water system is incompletely installed and does not function properly.
35. A 2011 report from EWCO staff also warned of the great tensions between sugar plantation investment and wildlife diversity: Cherie et al. (2011).
36. The presence of Chinese project workers has led to an upsurge of ivory hunting and illicit trade.
37. Information from Me'en informant, Addis Ababa, December 2016.
38. Last field information is of December 2016.
39. In 2016, alarming reports and pictures came out of local Suri people beaten up and put in slave-like chains. See: 'Ethiopia goes chain gang,' with photographs (http://thehornpost.com/ethiopia-goes-chain-gang/, accessed November 10, 2016). Human physical security in general has declined, due to the ongoing violent conflicts and enforced disarmament. Many locals (among them women and children) were killed by Kuraz Company drivers, and revenge actions were done in return, e.g., in November 2017: 13 highland drivers were killed by local people in a concerted attack.
40. Cp. Lewis (2015), Tickell (2015).
41. For example, the costs of the Omo Valley Kuraz sugar plantations outlay in 2011–2016 were *$3.6 bn* (Kamski 2016a, 568).

References

Abbink, J. (1998). Ritual and political forms of violent practice among the Suri of Southern Ethiopia. *Cahiers d'Etudes Africaines XXXVIII, 2–4*(150–152), 271–296.

Abbink, J. (2009a). The fate of the Suri: Conflict and group tension the Southwest Ethiopian frontier. In G. Schlee & E. E. Watson (Eds.), *Changing identifications and alliances in Northeast Africa* (pp. 35–51). Volume I: Ethiopia and Kenya, Oxford and New York: Berghahn Books.

Abbink, J. (2009b). Suri images: The return of exoticism and the commodification of an Ethiopian "tribe". *Cahiers d'Études Africaines, 49*(4), 196: 893–924.

Abbink, J. (2017a). Insecure food: Diet, eating and identity among the Ethiopian Suri people in the age of development. *African Study Monographs, 38*(3), 119–146.

Abbink, J. (2017b). Stemming the tide? The promise of environmental rehabilitation scenarios in Ethiopia. In W. van Beek, J. Damen, & D. Foeken (Eds.), *The face of Africa. Essays in honour of Ton Dietz* (pp. 185–198). Leiden: ASCL.

Addis Standard. (2017). *Analysis: The sour taste of sugar in Ethiopia. Corruption, incompetence and empty hope.* November 21, 2017. http://addisstandard.com/analysis-the-sour-taste-of-sugar-in-ethiopia-corruption-incompetence-and-empty-hope/. Accessed December 6, 2017.

Amborn, H. (2012). Ecocultural control of natural energy resources in Southern Ethiopia. *Aethiopica, 15,* 118–135.

Ashkenazi, M. (2012). Development is destruction, and other things you weren't told at school. In V. C. Franke & H. Dorff (Eds.), *Conflict management and "Whole of Government": Useful tools for U.S. National Security Strategy?* (pp. 91–126). Carlisle, PA: Strategic Studies Institute.

Asnake, K., & Fana, G. (2012). Discrepancies between traditional coping mechanisms to climate change and government intervention in South Omo, Ethiopia. In Mulugeta Gebrehiwot Berhe & Jean-Bosco Butera (Eds.), *Climate change and pastoralism: Traditional coping mechanisms and conflict in Africa* (pp. 132–152). Addis Ababa: Institute for Peace and Security Studies—University for Peace.

Avery, S. (2012). *Lake Turkana and the Lower Omo: Hydrological impacts of major dam and irrigation development (2nd Vol.).* Oxford: African Studies Centre, University of Oxford.

Avery, S. (2014). *What future for Lake Turkana.* Oxford: African Studies Centre, University of Oxford.

Barbier, E. B. (2014). Account for depreciation of natural capital. *Nature, 515,* 32–33.

Behnke, R., & Kerven, C. (2013). *Counting the costs: Replacing pastoralism with irrigated agriculture in the Awash Valley, North-Eastern Ethiopia* (Working Paper No. 4). London: International Institute for Environment and Development.

Belete, B., Assefa, M. M., & Yilma, S. (2013). GIS-based hydrological zones and soil geo-database of Ethiopia. *Catena, 104,* 21–31.

Berhanu, L., & Nejib, M. (2016). *Prosopis juliflora* in Gamo Gofa, Segen area and South Omo zones, Southern Ethiopia. *International Journal of Agricultural Science and Technology, 4*(1), 19–27.

Breu, T., Höggel, F. U., & Rueff, H. (2015). *Sustainable livestock production? Industrial agriculture versus pastoralism.* Bern: CDE (CDE Policy Brief, No. 7).

Brittain, M. W., & Clack, T. A. R. (2012). Pristine wilderness, participatory archaeology, and the custodianship of heritage in Mursiland. In T. Sternburg & L. Mols (Eds.), *Changing deserts: Integrating environments and people* (pp. 192–212). Cambridge: White Horse Press.

Brittain, M. W., Clack, T. A. R., & Bonet, J. S. (2013). Hybridity at the contact zone: Ethnoarchaeological perspectives from the Lower Omo Valley Ethiopia. In W. P. van Pelt (Ed.), *Archaeology and Cultural Mixture, Special Issue of Archaeological Review from Cambridge, 28* (1), 133–150.

Buffavand, L. (2016). 'The land does not like them': Contesting dispossession in cosmological terms in Mela, South-West Ethiopia. *Journal of Eastern African Studies, 10*(3), 476–493.

Buffavand, L. (2017). *Vanishing stones and the Hovering Giraffe: Identity, land and the divine in Mela, South-West Ethiopia* (PhD thesis, Martin–Luther Universität Halle-Wittenberg, Halle (Saale)).

Cardinale, B. J., et al. (2012). Biodiversity loss and its impact on humanity. *Nature, 486,* 59–67.

Carr, C. J. (2012). Humanitarian catastrophe and regional armed conflict brewing in the transborder region of Ethiopia, Kenya and South Sudan: The proposed gibe III dam in Ethiopia. African Resources Working Group (ARWG) Report (online).

Carr, C. J. (2017). *River basin development and human rights in Eastern Africa—A policy crossroads*. London: Springer (E-book).

Cherie, E., Derbe, D., & Girma, T. (2011). *Existing challenges: Plantation development versus wildlife conservation in the Omo-Tama-Mago complex.* Addis Ababa: Ethiopian Wildlife Conservation Authority (Report, 15 p.).

Clack, T., & Brittain, M. (2011a). Place-making, participative archaeologies and Mursi Megaliths: Some implications for aspects of pre- and proto-history in the Horn of Africa. *Journal of Eastern African Studies, 5*(1), 85–107.

Clack, T., & Brittain, M. (2011b). When climate changes: Megaliths, migrations and medicines in Mursiland. *Current World Archaeology, 46,* 32–39.

Clack, T., Brittain, M., & Turton, D. (2017). Oral histories and the impact of archaeological fieldwork encounters: Meeting Socrates on the Omo. *Journal of the Royal Anthropological Institute, 23,* 669–689.

Cox, D. (2014, May 24). The 'miracle' berry that could replace sugar. *The Atlantic Monthly.* http://www.theatlantic.com/health/archive/2014/05/can-miraculin-solve-the-global-obesity-epidemic/371657/?_ga=1.136550576.123960932.1401885673. Accessed June 5, 2017.

Dessalegn, R. (2014). The perils of development from above: Land deals in Ethiopia. *African Identities, 12*(1), 26–44.

Ellison, J. (2012). The intimate violence of political and economic change in Southern Ethiopia. *Comparative Studies in Society and History, 54*(1), 35–64.

EWCA. (2010). *The value of the Ethiopian protected area system: Message to policy makers*. Addis Ababa: Ethiopian Wildlife Conservation Authority (Leaflet).

Eyasu, E., & Feyera, A. (2010). *Putting pastoralists on the policy agenda: Land alienation in Southern Ethiopia*. London: IIED.

Eyasu, Y., et al. (2015). *Multi reservoir operation and challenges of the Omo River Basin: Part II: Potential assessment of flood based farming on lower Omo Ghibe Basin*. Addis Ababa: IGAD, Inland Water Resources Management Programme.

Fana, G. (2015). Securitisation of development in Ethiopia: The discourse and politics of developmentalism. *Review of African Political Economy, 41* (suppl.), S64–S74.

FDRE. (2005). *National biodiversity strategy and action plan* (p. 103). Addis Ababa: Federal Democratic Republic of Ethiopia, Institute of Biodiversity Conservation.

FDRE Policy Research Centre. (2017). *Large-scale government project management: A study of challenges and ideas on solutions*. Adds Ababa: FDRE Policy Research Centre, Section Good Governance and Capacity Building (327 p., in Amharic).

FDRE and UNDP. (2006). *Global environment facility, full project: Sustainable Development of the Protected Area System of Ethiopia (SDPASE). PIMS 494*. Addis Ababa: UNDP.

Fekadu, A. (2014). Politics of territoriality in Ethiopia: The case of the pastoral Gabra of Southern Ethiopia. *Ethiopian Journal of the Social Sciences and Humanities, 10*(2), 25–50.

Feron, B. (2008). *Surma, faces and bodies*. Bruxelles: La Renaissance du Livre.

Fong, C. (2015). *The scramble for water, land and oil in the Lower Omo Valley*. Berkeley, CA: International Rivers. www.internationalrivers.org/sites/default/files/attached-files/the_scramble_for_water_land_and_oil_in_the_lower_omo_valley.pdf. Accessed May 29, 2016.

Garretson, P. (1986). Vicious cycles: Ivory, slaves and arms on the new Maji frontier. In D. L. Donham & W. James (Eds.), *The southern Marches of imperial Ethiopia, essays in history and social anthropology* (pp. 196–218). Cambridge: Cambridge University Press.

Getachew, S. (2014). *The nexus of indigenous ecological knowledge, livelihood strategies and social institutions in Midland Gedeo Human-Environment Relations*. PhD dissertation in Social Anthropology. Addis Ababa: Addis Ababa University.

Giles, J. (2005a). Millennium group nails down the financial value of ecosystems. *Nature, 434*, 547.
Giles, J. (2005b). Ecology is key to effective aid, UN told. *Nature, 437*, 180.
Gil-Romera, G., Turton, D., & Sevilla-Callejo, M. (2011). Landscape change in the lower Omo Valley, southwestern Ethiopia: Burning patterns and woody encroachment in the savannah. *Journal of Eastern African Studies, 5*(1), 108–128.
Girke, F. (2013). *Homeland, boundary, resource: The collison of place-making projects on the lower Omo River, Ethiopia* (Working Paper No. 148, 24 p.). Halle (Saale): Max-Planck Institute for Social Anthropology.
Gross, K. (2016). Biodiversity and productivity entwined. *Nature, 529*, 293–294.
Hallman, B., & Olivera, R. (2015, April 16). Gold rush. How the World Bank is financing environmental destruction. *Huffington Post*. http://projects.huffingtonpost.com/worldbank-evicted-abandoned/how-worldbank-finances-environmental-destruction-peru?ncid=tweetlnkushpmg00000067. Accessed August 6, 2016.
Hendry, J. (2014). *Science and sustainability. Learning from indigenous wisdom*. London and New York: Palgrave Macmillan.
Hirsch, E., & O'Hanlon, M. (Eds.). (1995). *The anthropology of landscape: Perspectives on place and space*. Oxford: Clarendon Press.
HRW. (2012). *"What will happen if hunger comes?" Abuses against the indigenous people of Ethiopia's lower Omo Valley*. New York: Human Rights Watch.
HRW. (2017). *Ethiopia: Dams, plantations a threat to Kenyans. Statement*, February 14, 2017. https://www.hrw.org/news/2017/02/14/ethiopia-dams-plantations-threat-kenyans.
IFAD-Montpellier Panel. (2014). *No ordinary matter: Conserving, restoring and enhancing Africa's soils*. London: Agriculture for Impact, Imperial College London.
IGAD. (2016). *The IGAD regional biodiversity policy plan*. Addis Ababa: IGAD. https://static1.squarespace.com/static/5670f5fe2399a33f301331a2/t/57570d05ab48de200c8efd45/1465322773659/IGAD+Regional+Biodiversity+Policy+3+June.pdf. Accessed November 3, 2017.
IIED. (2014). *Biocultural heritage territories*. London: International Institute for Environment and Development.
ILO. (1991). *Indigenous and Tribal people convention, 1989 (no. 169) (27 June 1989)*. www2.ohchr.org/english/law/indigenous.htm. Accessed April 5, 2015.

Kamski, B. (2016a). The Kuraz Sugar Development Project (KSDP) in Ethiopia: Between 'sweet visions' and mounting challenges. *Journal of Eastern African Studies, 10*(3), 568–580.

Kamski, B. (2016b). *The Kuraz Sugar Development Project*. East Lansing, MI: Omo-Turkana Research Network (Briefing Note 1).

Keeley, J., ct al. (2013). *Large-scale land deals in Ethiopia: Scale, trends, features and outcomes to date*. London: International Institute for Environment and Development.

Korf, B., & Schetter, C. (2012). Räume des Ausnahmezustands. Carl Schmitts Raumphilosophie, frontiers und ungoverned territories. *Peripherie (Münster), 32*(126–127), 147–170.

Krätli, S., et al. (2013). Pastoralism: A critical asset for food security under global climate change. *Animal Frontiers, 3*(1), 42–50.

Lewis, K. (2015). *Ethiopia's Kwegu tribe in dire situation, reports say*. https://www.voanews.com/a/kwegu-tribe-water-dam-ethiopia-food-starving-government-resettlement/2719883.html. Accessed December 20, 2017.

Li, T. M. (2010). To make live or let die? Rural dispossession and the protection of surplus populations. *Antipode, 41*(S1), 66–93.

Lisanework, N., et al. (2010). Impact of parthenium hysterophorus on grazing land communities in Northeastern Ethiopia. *Weed Biology and Management, 10*, 143–152.

Low, S. M., & Lawrence-Zúñiga, D. (Eds.). (2003). *The anthropology of space and place. Locating culture*. Malden—Oxford and Victoria—Berlin: Blackwell.

Markakis, J. (2011). *Ethiopia: The last two frontiers*. Woodbridge, UK: James Currey.

Masood, E. (2015). Major biodiversity initiative needs support. *Nature, 518*, 7.

Meles, Z. (2011, January 25). *Speech at the 13th annual pastoralist day of Ethiopia. Jinka, South Omo, Ethiopia*. www.mursi.org/pdf/Meles%20Jinka%20speech.pdf. Accessed March 20, 2013.

Newbold, T., et al. (2015). Global effects of land use on local terrestrial biodiversity. *Nature, 520*, 45–50.

Newsome, M. (2015). The people pushed out of Ethiopia's fertile farmland. *BBC News*. http://www.bbc.com/news/magazine=30623571. Accessed June 15, 2015.

Oakland Institute. (2013a). *Understanding land investment deals in* Africa. Ignoring abuse in Ethiopia—DFiD and USAID in the lower Omo Valley (Report). Oakland, CA: Oakland Institute.

Oakland Institute. (2013b). *Omo: Local tribes under threat. A Field report from the Omo Valley, Ethiopia*. Oakland, CA: Oakland Institute.

Oakland Institute. (2014a). *Engineering ethnic conflict: The toll of Ethiopia's plantation development on suri people*. Oakland, CA: Oakland Institute.

Oakland Institute. (2014b). *Willful blindness. How word bank's country rankings impoverish smallholder farmers*. Oakland, CA: Oakland Institute.

Olisarali, O., & LaTosky, S. (2015). Pastoralists do plan! Experiences of Mursi land use planning, South Omo, Ethiopia. *Making Rangelands Secure, 6*, 4–5.

Pardo de Santayana, M., & Manuel, J. M. (2015). The benefits of traditional knowledge. *Nature, 418*, 487–488.

Perry, M. (2015). Dismantling the Omo Valley. *Sustainable food trust*. February 13, 2015. http://sustainablefoodtrust.org/articles/land-grabbing-omo-valley/. Accessed November 20, 2015.

Polanyi, K. (2001/1944). *The great transformation. The political and economic origins of our time* (2nd ed.) (Foreword by J. Stiglitz. Introduction by F. Block). Boston: Beacon Press.

Rettberg, S., & Müller-Mahn, D. (2012). Human-environment interactions: The invasion of *Prosopis juliflora* in the drylands of Northeast Ethiopia. In L. Mol & T. Sternberg (Eds.), *Changing deserts* (pp. 297–316). Cambridge: Whitehorse Press.

Ricketts, T. H., et al. (2010). Indigenous lands, protected areas, and slowing climate change. *PLoS Biology, 8*(3), e1000331.

Sahlins, M. (2000). 'Sentimental pessimism' and ethnographic experience: Why culture is not a disappearing 'object'. In L. Daston (Ed.), *Biographies of scientific objects* (pp. 158–202). Chicago and London: University of Chicago Press.

Sassen, S. (2014). *Expulsions: Brutality and complexity in the global economy*. Cambridge, MA: Harvard University Press/The Belknap Press.

Scott, J. C. (1998). *Seeing like a state: How certain schemes to improve the human condition have failed*. New Haven, London: Yale University Press.

Seyoum, M. (2015). The nexus between land-grabbing, livelihood insecurity and conflict in Ethiopia: The case of Majang in the Gambella Region. *African Peace and Conflict Journal, 8*(1), 53–67.

Silvester, H. (2009). *Éthiopie. Les Peuples de l'Omo*. Paris: Éditions de La Martinière.

Smallteacher, R. (2013, July 18). *Ethiopian sugar alleged to destroy pastoral communities of lower Omo*. http://www.corpwatch.org/article.php?id=15856. Accessed June 15, 2015.

Stevenson, E. G. J., & Buffavand L. (2018). 'Do our bodies know their ways?' Villagization, food insecurity, and ill-being in Ethiopia's Lower Omo Valley. *African Studies Review, 61*(1) (forthcoming).

Temperley, M. (2015, May 24). Picture story: How photographing the Omo Valley people changed their lives. *The Guardian*.

Tewolde, W., & Fana, G. (2014). Socio-political and conflict implications of sugar development in Salamago Wereda, Ethiopia. In Mulugeta Gebrehiwot Berhe (Ed.), *A delicate balance: Land use, minority rights and social stability in the horn of Africa* (pp. 117–143). Addis Ababa: Institute for Peace and Security Studies.

Tickell, O. (2015, March 13). Ethiopia: Kwegu tribe starves, victims of dam and land grabs. *The Ecologist*. https://theecologist.org/2015/mar/13/ethiopia-kwegu-tribe-starves-victims-dam-and-land-grabs. Accessed December 20, 2017.

Tsegaye, M. S. (2016). Large-scale land acquisitions, state authority and indigenous local communities: Insights from Ethiopia. *Third World Quarterly, 38*(3), 698–716.

Turton, D. (2005). The meaning of place in a world of movement: Lessons from long-term field research in Southern Ethiopia. *Journal of Refugee Studies, 18*(3), 258–280.

Turton, D. (2011). Wilderness, wasteland or home? Three ways of imagining the lower Omo Valley. *Journal of Eastern African Studies, 5*(1), 158–176.

Turton, D. (2012). Ethiopia: Concerns about Gibe 3 Dam. *Pambazuka News 568*. http://pambazuka.org/en/category/features/79590. Accessed April 4, 2014.

UNESCO. (2015). *Mission report—Rapport de Mission. Report on the reactive monitoring mission: Gibe III Dam and Kuraz Sugar Plantation (Ethiopia) for Lake Turkana national parks world heritage property (Kenya), From 3 to 7 April 2015*. Paris: UNESCO (35 p.).

UNESCO. (2017). *Lower Valley of the Omo (Ethiopia) (C17). Decision: 41COM 7B.68*. Paris: UNESCO. http://whc.unesco.org/en/decision/7068. Accessed December 10, 2017.

Vidal, J. (2015, September 3). EU diplomats reveal devastating impact of Ethiopia dam project on remote tribes. *The Guardian*.

Wagstaff, Q. A. (2015). *Development, cultural hegemonism and conflict generation in Southwest Ethiopia: Agro-pastoralists in trouble*. Bordeaux: Université Montesquieu (SciencesPo—LAM, Observatoire des Enjeux Politiques et Sécuritaires dans la Corne de l'Afrique. Online), 46 p.

WWF. (2004). *Sugar and the environment. Encouraging better management practices in sugar production*. Zeist: World Wildlife Fund.

7

Cameroon's Western Region: Environmental Disaster in the Making?

Moses K. Tesi

Introduction

Cameroon is often referred to as 'Africa in miniature' in reference to the fact that the country contains all the major geographic features found in various parts of Africa within its territory. These include Equatorial forests, savannah, and Sahel vegetations. The country's great variety of landforms, that begin with lowlands around the Atlantic Ocean give way to mountains, highlands, and plateaus northwards, and end with plains as one approaches Lake Chad, lends credibility to the nickname.

The designation of Cameroon as Africa in miniature could also be viewed in terms of its environmental challenges. Such challenges involve both 'green' and 'brown' environmental issues in the sense that the environmental challenges that are present in various parts of Africa are all present in Cameroon as well. Among them are deforestation,

M. K. Tesi (✉)
Department of Political Science and International Relations,
Middle Tennessee State University, Murfreesboro, TN, USA

desertification, erosion, issues of biodiversity decline, urban sprawl, and pollution. As is the case with many African countries, for decades the attention accorded to environmental issues by Cameroon's leadership primarily centered on the 'green elements' of the environmental landscape—deforestation, desertification, biodiversity threats, and related human activities that drive them (see Tesi 2001). The Cameroon leadership's attention to green environmental concerns, however, was concentrated only in particular regions of the country: the South, Southwest, the Coast, the East, Center, and the North (see Table 7.2).[1] Although brown environmental issues—among them carbon emissions, industrial waste disposal, urbanization and urban waste disposal, water and air pollution—have been growing during the past three decades in Africa and in Cameroon in particular, it is mainly since the mid-2000s that they started receiving serious attention from policy makers.[2] This is partly because Cameroon and other African countries have significant forest and desert ecosystems with a wider regional impact and are therefore indispensable for a global solution on biodiversity, deforestation, or desertification issues. The employment and revenue generating power of the continent's savannah and forest zones in terms of wildlife, timber, and forest tourism or timber was and remains a major factor in the interest such ecosystems generate (cf. Tesi 2001). In the case of Cameroon, timber exports ranked third of the country's exports and contributed US $380 million in earnings to the national treasury or 16% of export earnings in 2003 (see Bele et al. 2011). And in 2006, over 163,000 people were employed in the forestry sector.[3] On the other hand, the brown environmental issues are viewed as issues whose resolution depend more on the big emitters as China, the USA, India, and Brazil than African countries. Like green environmental challenges, policy attention on brown environmental issues in Cameroon is also concentrated in particular parts of the country—the industrial belt of the Coastal Region (Douala, Edea) and the Central Plateau (Yaounde–Mbalmayo).[4] A review of the country's environmental policy reveals that Cameroon's strategy is built on green and brown environmental foundations (see Rainbow Environment Consult 2007). Policy focus on the two is in turn driven by external constituencies and factors absent which is questionable whether the Cameroonian state with its priority centered on development at all cost would have committed to seriously addressing the environmental problems.[5] In focusing on green and brown environmental issues, the Cameroonian state's approach has been much too narrow, a phenomenon that reflects the influence of its colonial past and international activism—intergovernmental, non-governmental, and bilateral. Such external influence plus the government's own low priority on the environment made the latter to mainly emphasize the areas of interest to the external actors.

This chapter calls attention to a particular ecological zone of Cameroon that does not contain environmental resources of particular interest and urgency to external environmental advocates, but that is an evolving environmental time bomb. There is a growing environmental crisis in Cameroon's Western Region, and in particular, the Bamileke districts of that region. The chapter analyzes the origins of Cameroon's environmental protection policy, and where if at all the Western Region fits in that protection, reviews the region's main economic activities, land use practices, urban growth, and the country's development policies. I look at how they relate to fostering prosperity and to whether a sustainable environment is furthered or not. The chapter is set up as follows: (a) brief background of the region; (b) an examination of the various ecological factors whose interactions characterize the region's environment; (c) a discussion of the impact of the interactions; (d) a discussion of what is being done to address them; and (e) a discussion of various options that could be pursued to address and remedy the situation and the challenges associated with each.

The Natural Features of Cameroon's Western Region

The Western Region of Cameroon is one of the ten administrative regions of the country (see Fig. 7.1). The region is in the west-central part of the country[6] and is bordered by five other regions—Adamawa and the Northwest regions in the North, the Center region to the East, Littoral to the South, and Southwest to the West. The region lies on the

Western Plateau, with elevations that rise to 2500 m around the central location of Bafoussam, the region's chief city and capital. Its topography is marked by hills, escarpments, meandering valleys, and streams that make it one of the areas of magnificent natural beauty in the country. The Bamboutos Mountain, located in Babajou near Mbouda at the Northwest part of the region, is the highest point at 2740 m. From the central part of the plateau where Bafoussam is, the region then begins to descend to the low-lying regions of Littoral to the Southwest at the upper banks of the Nkam River, and then to the Mbam River valley in the Center Region to the East, and Southeast. Three main rivers drain the region. The Nkam River that begins from the Bamboutos Mountain drains it to the West and is also the headwater of the Wouri. The Noun River is one of the tributaries of the Sanaga River and forms the boundary between the Bamouns, indigenes of the Noun Division, and the seven Bamileke departments of the region. This river drains the region to the East. A dam on the Noun at Bamindjing for regulating the waters of the Sanaga at the hydroelectric plant in Edea has created a huge man-made lake extending to the Ndop plains in the Northwest Region. East of Noun Division is the third river, the Mbam. It forms the boundary between the Western Region and the Center. In addition to the rivers, the region is also home to lakes and waterfalls. The climate is generally equatorial, with minor variations. The region's high elevation provides it with cool temperatures that average 21–22 degrees. It experiences two seasons: The dry season that begins in November and ends in April, and the rainy season that begins in May and ends in October. Average annual rainfall is between 1800 and 2000 mm.

Population Pressure

The relationship between population and the environment is well-established in the literature, as is that between population, agriculture, and the ecosystem (see Cleaver and Schreiber 2001). Clearly, because of the 'fixed' nature of land, increases in the agricultural population of an area, whether the result of high birth rates or because of in-migration, have the effect of reducing the average size of available land per household, without notable increases in productivity. That is the predicament of the Western Region of Cameroon. The Western Region is the smallest of Cameroon's ten regions, with a size of 13,892 square km and a total population of 1,906,831 million people in 2015 and a population density of 137 per square km (as shown in Table 7.1). This compared unfavorably with 8 inhabitants per square km in the East, 16 in the South, and 59 in the Center. The numbers, however, do not tell us everything. Of the area of 13,892 square km that constitutes the Western Region, less than half, that is 6196 square km, is Bamileke country. Bamileke country is the part of the region that is west of the Noun River and lies predominantly on the hills and the plateau. Of the region's 1.9 million inhabitants in 2015, about 1.5 lived in the seven Bamileke Divisions. The population density in Bamileke country, based on the census numbers, was much higher than the ratio for the entire region, around 242 people per square km.

Table 7.1 Cameroon's population by region, 2015

Region	Population	Area (square km)	Density
Adamawa	1,183,551	63,701	19
Center	4,098,592	68,953	59
East	832,869	109,002	8
Far North	3,945,168	34,263	115
Littoral	3,309,558	20,248	163
North	2,410,936	66,090	36
Northwest	1,950,667	17,300	113
South	745,198	47,191	16
Southwest	1,534,232	25,410	60
West	1,906,831	13,892	137

Source Bureau Central des Recensements et des Etudes de Population, Yaounde 2015

Economic Activities in the Region

The principal economic sector in the Western Region of Cameroon is agriculture. But agriculture in the region is by smallholders, unlike the type of corporate plantations that dominate agriculture in the Littoral, Southwest, or Center and South Regions of the country. Another distinction of the region's agricultural economy is that many farmers today have given up cultivating coffee, the main cash crop that dominated the region's economy in the 1960s, 1970s, and 1980s, and focused instead on the production of food crops (Jiotsa et al. 2015). This was due to the drastic fall in the price of coffee and to the economic crisis that the country faced in the 1980s and 1990s.[7] Even without the economic crisis, the region has always been a leader in food production, next to producing coffee.[8] This coffee production was one of the agricultural activities basically imposed on the people of the region and other regions of Cameroon by the authorities, first the colonial and later the post-independence government (cf. Teretta 2014). The emphasis on food production—especially corn, plantains, cassava, tomatoes, spices, groundnuts, beans, potatoes, rice in some parts, and vegetables—has increased food production for the local markets in the big cities of Douala and Yaounde, and for export to CEMAC countries (Gabon, Republic of the Congo, Equatorial Guinea, and even Central Africa Republic).[9]

Economic activities built around small hold farming as practiced in the Western Region present serious challenges as to their sustainability. First, earnings in the region for over 95% of the population in one way or another are derived from working on the farms. Some civil servants and small business owners still cultivate crops on the side. This is quite obvious during planting and harvesting seasons of the major food crops in the region (corn, groundnuts, and beans). When one thinks of cities, the image comes to mind of large population centers bustling with peoples engaged in a variety of differentiated livelihood activities. This is certainly true also in cities in the Western Region and the entire country. But when the first rains fall, marking the beginning of the planting season, one notices a major difference in the big cities of the Western Region.[10] Many people operating small businesses such as artisanry, drinking houses, tailoring, small shops, and even taxi drivers, start the day (except on market days) by going to plant on their farms, returning in the afternoon to engage in their regular activities.[11] In addition to small business owners, artisans, etc., doing it as secondary activity, many other city dwellers in the Western Region are actually engaged in full-time farming for a living.[12] In Mbouda, for example, during the month of March and April, the major roads to the outskirts in the surrounding arrondissement of Galim, Babajou, Bacham, and Bafounda are usually crowded with people leaving the city for their farms.[13] This same phenomenon was also observed during the peak harvest months for corn and groundnuts, in July and August.[14] Similar observations can be made in the case of other cities in the region and their surrounding areas—especially in the Bamileke departments, e.g., as one leaves Mbouda to Bafoussam or to Dschang, Bafoussam to Bafang, or Bangante.[15] The surroundings of these cities, mainly hills intersected by valleys, are mostly under cultivation. What also stands out is the proximity of the cities of the region to each other (See Table 7.2). Farming is so crucial to peoples' lives that, as an intercity minibus driver on my way from Mbouda to Bafoussam told me: 'There is no piece of land that has not been laid claim to by someone and very little difference exists between city and rural areas when farming is concerned.'[16] Other passengers in the minibus quickly jumped into the conversation saying that even if a piece of land lays barren for decades, no one should tamper with it because the owner will emerge once someone tries occupying it.[17]

The Nexus: Limited Land, High Population Density, Intensive Farming, Urbanization, and Ecological Stress

The region's high population density is mainly due to local increase in a small geographic space and only marginally to immigration from other regions of Cameroon.[18] The impact of such increase is visible in all areas of socioeconomic activity. High population density has placed immense pressure on the ecosystem due to the limited arable land available in proportion to the number of people engaged in farming. Such pressure is more severe in the seven Bamileke Divisions, as noted above, than in the Noun Division, and has contributed to the expansion of farmlands to ecologically delicate areas on slopes, hilltops, river banks, wetlands, and whatever is left of forest ecosystems discussed above.[19] Population pressure has also contributed to farming activities being carried out closer to the frontiers with other villages and towns in the region without respecting the empty space and natural vegetation cover that traditionally used to separate one village from another. Another practice that is commonly carried out in the region is the cultivation of almost any patch of land available, whether it is in the city or its vicinity. Despite over-cultivation of farmlands—a departure from the traditional technique of shifting cultivation that once was practiced in the region, modern agricultural techniques (fertilizer use, irrigation, multiple cropping, crop rotation, pesticide use, and multiple growing seasons) have helped to increase yields, even as problems associated with land scarcity looms. In fact, one of the reasons for the success of agriculture in the Western Region is due to the embrace of the use of such new techniques. That notwithstanding, the challenge remains that with a fixed, circumscribed geographic space dominated by agriculture the issue of addressing other human activity-related impacts on the environment such as dwellings, rural sprawl, and administratively fueled urbanization, beside the economics of growth in agricultural productivity, need to be urgently addressed.

Too many people chasing limited land that has led to the expansion of agricultural activities to marginal and delicate ecosystems have been significantly helped by other factors to make the Western Region emerge as an environmental time bomb. One such factor is the government's own actions in creating administrative centers known as divisions and subdivisions without careful thoughts put to it regarding the effects on the ecology. This was more so in the Bamileke part of the region than in the Bamoun part. Administrative divisions and subdivisions were often created to allegedly bring 'government services closer to the population.'[20] But the government also used its power to create the new divisions as political rewards to its middlemen for their work in 'delivering the votes' from their communities during the critical elections of the 1990s. This exacerbated the urbanization problem in the region. There are eight administrative divisions in the Western Region, seven of which are in Bamileke country. Each of these divisions is organized into subdivisions. And there are 37 subdivisions, 29 of which are in Bamileke country.[21] The creation of divisions and subdivisions has contributed to the growth of these places into urban centers because people move there from rural areas, usually from the same region, in search of better economic opportunities. This has had the effect of contributing to the emergence of many new urban areas where none existed before.[22] These towns moreover emerged not in the traditional way, due to new industries that came with jobs, but out of anticipation by the people moving in that with government offices would come business and jobs.[23] But these jobs have not really materialized. Associated with the creation of administrative divisions has been the tendency of the government to expropriate large acreages of land to build office buildings and other infrastructures for the public services to be delivered. But not only do the officials who staff the services come from outside the area, even those who work on many such projects also came from outside. So very little in terms of economic benefits accrues to the local population of the area, especially the subdivisions. The net effect has thus tended to be that many inhabitants of the areas so affected end up working in the informal economy as day laborer, taxi drivers, etc. Meanwhile, the lack of proper amenities such as proper sanitation, housing, sewage, and roads also turn many of the new urban centers into shanty towns or slums, even the smaller ones.

The government's role in the phenomenon of urbanization in the region has produced a new economic dynamic, but also had a major effect on the environment. Owners of land expropriated by the government often lost access to their land. This also meant a loss for the environment, because crucial ecological regulatory functions of smallholder agriculture (including fallow cycles) were undermined after expropriation and following new intensive commercial exploitation or urban use (Tankou et al. 2013). Cameroon land law stipulates that land that 'does not have a title' could be expropriated without compensation by the government for use, because all land in principle is government land.[24] The government is required to pay the owner of land that was developed as compensation for the value of what is on it if he/she has a land title. But most land in rural and urban areas is governed by the tradition and customs of the indigenous people of the area in question. Landowners do not usually have a 'land title,' as that is a new phenomenon. Even people whose land the government is required to pay compensation for do not always get the payment without spending both time and effort pursuing the government for months and sometimes for years.[25]

Another way government actions contributed to rapid urbanization in the region was the strategy it adopted in the early 1960s in its fight against the *Union des Populations du Cameroun* (UPC) insurgency.[26] The government completely changed the dwelling pattern in the region by mandating that people should move from their dwellings in remote parts of their villages and regroup along the roadsides.[27] The objective was to better trace insurgents. But while the government got what it wanted, the policy also congregated people in ways that made it easier for their areas to become urbanized faster. For example, the population of Bafoussam that was 7000 in 1956 jumped to 20,000 in 1960. Bafang's population jumped from 5200 in 1957 to 11,500 in 1962 (Champaud 1983, 84).

The urban population makes up 52% of Cameroon's total population. Although the largest cities in the country, Douala and Yaounde, are not in the Western Region, five of the twenty large cities are in the Western Region: Bafoussam, Dschang, Foumban, Mbouda, and Foumbot. The Western Region also ranks third in the number of cities in the country with a population of 15,000 and above, after the Center and Far North.[28] But unlike other regions, the small size of the region makes distances from one division or city to another to be very short. Consequently, people can access government services with ease if investments are put on building better roads and providing peoples' basic social and economic needs. The longest distance from one Department's chief town to that of another is between Bafang and Foumban, which is 128 km (80 miles). All the other departments are very close to each other. The regional headquarters Bafoussam is located at the center of the region. This means that with better roads, the people of the region would be able to access government services easily, meaning that there is basically no need for creating so many administrative divisions or subdivisions. Because these divisional headquarters and their subdivisions are so close to each other, with continued growth it will not be long before some of them, in particular Bafoussam Mbouda-Banjoun-Baham-Foumbot, will expand and connect to each other, creating a massive urban sprawl area. Under present circumstances, with continued high population density and growth, agricultural lands will further shrink, and green spaces will decrease due to the ever-growing connectedness of the divisions.

Part of the problem with urban expansion is its sprawl-like character. This is due to the absence of housing codes to regulate home construction, and the result of a building culture that often emphasize large sprawling homes, and not multi-storey homes. The house-building is also on fragile lands, often with mud bricks and without indoor plumbing. Modern buildings in the region tend not to always follow construction codes, and at times do not allow adequate distance from the road or from neighboring homes. Small homes built up, with more storeys, would take less space and land, and conserve unused land for other purposes. The large, spreading houses consume the limited land available and exacerbate the land problems.

Finally, the dependence of the region on agriculture has made its environmental prospects very uncertain. Its population continues to grow, although at a lower rate, and available land for farming has declined per household. This means that agricultural yields will not be growing as much they should for employment and feeding the cities. Industry will be needed to absorb the workforce of the region, or the population will need to migrate to other parts of the country either where there is reputed to be 'abundant land' to pursue farming activities, or where there are industries for them to seek non-agricultural employment.

Past Efforts to Deal with the Issue

The challenge of too many people chasing very little land was recognized as far back as the colonial era, even before the emergence of modern environmental concerns. When the country gained independence in 1960, the new government also became concerned about the population pressure in the region. The French colonial government had encouraged various Bamileke groups to migrate and settle in the Eastern parts of the Noun River in the Noun Division, where there was ample farmland. There was no policy or even awareness of any need to reduce their population pressure (cf. Teretta 2014, 52–54). Large numbers were also attracted to the Mungo Region South of Bamileke country to work as laborers on plantations owned by French settlers. Many who migrated to work on European plantations in the Mungo Region in towns like Melong, Nkongsamba, Manjo, Mbanga, Njombe, Loum, and Pemja ended up acquiring their own farms (cf. Dongmo 1981, 118, 231; Terretta 2014, Chapter 2; Barbier 1977; Champaud and Gendreau 1983). Others moved to Douala and Yaounde, where they worked in various unskilled activities for Europeans, in the port and in the few industries that were beginning to emerge. Meanwhile, others worked in informal activities, learnt trades, and eventually became the largest of the various African groups in the cities (see Johnson 1970, 50–52; LeVine 1964, 64).

After independence the Ahidjo government devised a resettlement project in Nkondjock between 1965 and 1970, along the newly constructed Bafang Yabassi road, as the French had done on the Bamoun side of the Noun, giving them free land to farm on.[29] But the Bamileke working outside their homelands still maintained links back in the Western Region, and often upon retirement returned there. Thus, the pressure on land in the region did not end with out-migration. Out-migration merely decreased it as those who lived outside the area did not need land to farm on but needed it to build on.[30]

Another causal factor of the pressure on land and the ecology was the failure of the colonial government to establish many industries in Cameroon, and those that they established were mainly in the Douala, Edea, Yaounde, Limbe, and Garoua areas. None was in the Western Region. Under Cameroonian independence rule after 1960, no significant industrial activities were started in the region. Up to 75% of the country's industrial activities, according to one source, are in Douala[31] and the majority of the remainder are located in Edea, Yaounde, Garoua, Kribi, or Limbe.[32] In the Western Region, the only industrial activity of note was carried out by UCCAO, a cooperative of coffee producers.[33] UCCAO processed some of the coffee produced there and sold it on the local market and to CEMAC (the Economic Community of Central African States). Brewery and soap making are the only other industrial activities that now operate in the region, but they are not big enough to provide as a significant number of jobs.[34] Thus, many of those in the region who do not engage in agriculture for a living are those who work in schools and Universities, branches of banks, various government agencies, hospitals, the police, military, or are small business owners involved in trading or are artisans. These do not come close to 5% of the region's population. So, until a significant ratio of the population is diverted from agriculture to other non-environmentally tasking activities, the combined forces of population, agricultural expansion, limited land, and human habitation would generate a severe crisis in the Western Region. The issues discussed above make the Western Region an environmental challenge that requires urgent action before it is too late.

Cameroon's Environmental Policy and the Western Region

Concern for the environment in Cameroon goes back to the colonial era. Forest reserves and wildlife protection were at the forefront of the first environmental policies during the colonial era: in 1932, the Mozogo-Gokoro National Park was created in today's Far North Region and in 1934 the Waza wildlife Reserve. At independence in 1960, there were ten reserves in Cameroon. From 1960 to 1980, five more were added. Six more were created in 2000, bringing the total to twenty-one, as shown in Table 7.3. Of that number, only one was in the Western Region. Post-independence Cameroon merely continued with the colonial strategy of basing the classification of the protected area reserves on their role as vacation areas. This made the newer protected areas to be either in the heavily forested regions or in the savannah regions of the country, whose vegetation made them to be habitat for many of the country's wildlife (See Table 7.3). In other words, initially, Cameroon's environmental strategy did not deal with the range of issues related to each of the country's *ecosystems*. Ecosystems in the Western Region were ignored as such and not valued for their wider environmental role (Tieguhong and Betti 2008). Having no wildlife to speak of, they did not possess nor present the drama that wildlife of the savannah or the thick forests had (Table 7.3).

Table 7.2 Distances from Bafoussam capital of the Western Region to major towns in the region in kms

Capital of Western Region	Major town	Distance in km
Bafoussam	Dschang	52
Id.	Mbouda	29
Id.	Foumban	70
Id.	Baham	21
Id.	Bafang	58
Id.	Banjoun	15
Id.	Foumbot	27
Id.	Bangante	49

Table 7.3 Cameroon's protected areas

Protected reserves	Hectares	Regions	Created	Made into National Park
Waza National Park	170,000	Far North	1932	1968
Kalamalouse National Park	4500	Far North	1947	1972
Mozogo-Gokoro	1400	Far North	1932	1968
Benue National Park	180,000	North	1932	1968
Faro National Park	330,000	North	1932	1968
Korup National Park	126,000	Southwest	1962	1982
Dja Reserve	526,000	East	1950	–
Douala-Edea Wildlife Reserve	160,000	Littoral	1932	–
Bouba-Ndjda	220,000	North	1947	1968
Lobeke Wildlife Park	43,000	East	–	2000
Campo Wildlife Park	271,000	South	1932	2000
Kalfou Wildlife Park	4000	Far North	1933	–
Lake Ossa Forest Reserve	4000	Littoral	1968	–
Kimbi Forest Reserve	5600	Northwest	1964	–
Santchou Forest Reserve	7000	West	1968	–
Mbi Crater	400	Northwest	1964	–
Mengame Game Sanctuary	17,500	South	–	2000
Mbayang Mbo Game Sanctuary	Southwest	–	–	2000
Mbam et Djerem	416,512	Center/Adamawa	–	2000
Boumba Bek	210,000	East	–	2000
Lobeke	428,000	East	–	2000

Source Tchindjang et al. 2005

It was not until the 1992 Earth Summit in Rio that the view of the environment began to change to include elements beyond forests and wildlife conservation. Of the three conventions signed in Rio, the Convention on Biodiversity, Desertification, and Climate Change, the one that would target the Western Region would be that on desertification if liberally interpreted and that on climate change on solely ecological factors. This notwithstanding that various works and UN conferences had sounded the alarm on the many elements that needed to be considered regarding environmental issues (cf. Meadows 1972; World Commission on the Environment and Development 1987). Nonetheless, the Rio Summit was a turning point in the country's thinking on the environment and its relationship to human activities. Cameroon, in the efforts to meet the requirements of the conventions agreed to at Rio, created a separate Ministry on the Environment to better develop policies and create an institutional basis for dealing with the issues involved.[35] Until then, environmental issues fell under separate departments in the Ministries of Agriculture and Tourism (Mengang 1998). The activities associated with agriculture itself were not viewed to be related to the environment. Fertilizer use, for example, was viewed less on its potential to saturate the soil, sip into the water table, and contaminate groundwater with chemicals, than on its expectation to generate higher crop yields.

Later, Ministerial changes created two new ministries out of the Ministry of the Environment. These were the Ministry of Forests and Wildlife and the Ministry of the Environment. The Ministry of Forests and Wildlife dealt with sectoral issues regarding forests and fauna management while the Ministry of the Environment was supposed to draw up the government's overall environmental strategy and serve as the watchdog on environmental regulation. The new setup brought about the creation of a National Environmental Management Plan (NEMP) in 1996. Despite that, the government remained slow in embracing environmental challenges fully. It was not until 2004 that it ratified the United Nations Framework Convention on Climate Change (UNFCCC) negotiated at Rio in 1992. The government did not, however, develop its own environmental policy document as required by UNFCCC to deal with the key areas of focus—biodiversity, sustainable forest management, land degradation, conservation, and desertification. Instead, it used its Poverty Reduction Strategy Paper (PRSP) that it had negotiated with the World Bank and the International Monetary Fund (IMF) in 2003 as a condition for receiving debt relief under the Heavily Indebted Poor Countries Initiative (HIPCI), as its strategy document on the environment. But analysis by the International Institute for Sustainable Development (IISD 2004) notes that the PRSP did not emphasize the environment within its framework, even though the two were related given the PRSP's concerns with poverty reduction, *sustainable* economic growth, increased efficiency in public expenditures, improvement in governance, etc.[36] The Millennium Development Goal (MDG) adopted in 2000 with the goal of cutting global poverty by half was another source of external pressure on Cameroon to pursue environmentally sound policies more seriously than it had done so far. Such external pressure came with financial support if the country would take advantage of it to assess the impact of environmental factors, come up with mitigation measures, and devise adaptation strategies to cope with the problems, and then communicate this to the Conference of the Parties (COP) as required by the UNFCCC to get necessary financing (Norrington-Davies 2011). Cameroon's first communication on its greenhouse gas (GHG) emissions to the COP was in 2005, in which it reported having emitted 43 million tons of GHG at the 1994 baseline, made up of 55.9% CO_2, 25.3% CH_4, and 18.8% N_2O (Bele et al. 2011, 379). But while the government's inventory of the causes of the emissions focused on agricultural activities such as slash and burn, wood extraction for biomass use and land use changes as being responsible, the mitigation measures for which it sought financing focused on reforestation and sustainable forest management, reduction in energy consumption, and investments in renewable energy (Bele, ibid.) Little to no attention was given to possible mitigation strategies that would deal with agriculture, land use, or promoting industrial activities that would (while increasing employment activities for a larger number of people) generate high incomes, and at the same time eliminate pressure on the land. Consequently, external funding for environmental issues would be channeled mainly to the forestry sector. This from the onset excluded the Western Region, as there were neither forests nor wildlife there (see Norrington-Davies 2011). The absence of assessing impact and adaptation mechanisms also meant that neither the Western Region nor the country as a whole would have a thorough understanding of the consequences of environmental impact, or of the possibilities that such knowledge could lead to in terms of adaptation strategies. Even more, the government had not yet submitted a Reduction in Emission from Deforestation and Degradation (REDD), or a National Adaptation Plan of Action (NAPA) to the UNFCCC by 2015, meaning that it was not able to receive funding for projects (Kengoum and Tiani 2013). It was, however, working with the United Nations Environmental Program (UNEP) to develop a Readiness Preparation Plan (RPP).[37] Even so, the emphasis of the government in its 2012 and 2013 RPP was to reduce activities related to slash and burn agriculture and biomass.[38]

The government's attitude on environmental issues has been more hesitant and mixed rather than firm and ambitious. As Norrington-Davies notes, the government's focus is more on achieving its development goals under its Agenda 2035, and would do what it can to realize it, especially without the constraints that being monitored under the UNFCCC means. It therefore sees climate change and environmental issues as an 'inconvenience,' especially when external funding that comes with its commitment under UNFCCC is far less than what it earns through its less environmentally economic restrictive activities.[39] Where such benefits have been huge, it has worked hard to seize the opportunity. This was the case on the Trans-Cameroon Pipeline and the Kribi Deep Seaport projects, the massive projects to construct another pipeline from Cameroon's port city of Kribi to Chad and a deep seaport in Kribi.[40] The two projects required major environmental impact studies. Both outside pressure from external NGOs and the World Bank ensured that these were undertaken to their satisfaction. But, as discussed earlier, environmental concerns in the Western Region of Cameroon did not have such strong advocates, and for many years, the challenges that await the region have not been part of the agenda of various agencies involved in global environmental issues.

Without external forces that can withhold funding or pressure global leaders to withhold such funding, as sometimes happens with groups like the World Wide Fund for Nature, the World Conservation Union, the World Resources Institute, and others, the government of

Cameroon has felt little pressure to deal with the severe ecological stress in the Western Region. It has not done much to benefit regions without forests or wildlife ecosystems. The Western Region has missed out on the resources of the Ministry of Forest and Wildlife or that of the Ministry of Tourism as related to the touristic aspects of Forest Reserves or National Parks. The region's only main institution on environmental matters has been the Ministry of Agriculture, because of the Western Region's strong agricultural base. Other major Ministries with some jurisdiction over environmental issues of interest to the region are the Ministries of Urban Affairs and Health. These are of interest on issues related to urban pollution, sewage and waste disposal, and sanitary and public health issues. But these are issues to which less attention was paid during the country's economic crisis of the 1980s and 1990s. It was only in the 2000s that some initiatives were taken regarding urban waste disposal. Other sanitary issues including sewage, public health, faulty or lacking indoor plumbing, or dirty urban streets continued to pose major problems in the region's urban centers. Part of the reason for neglect was political, in that the region had until the mid- to late 2000s been a hotbed for the opposition, and the government used that against it in the distribution of public goods (Tesi 2018). Also, because the region did not have any industries considered to pollute it, it also missed out on external funding destined for cleaning up polluted urban centers (see Norrington-Davies 2011; Global Environment Facility: Country Portfolio Evaluation – Cameroon 1992–2007).

Going Forward

The environmental challenge in the Western Region of Cameroon is a classic example of what ecologists talk about: a challenge associated with a highly rising population, compressed in a very small space and whose activities impact the normal operation of the ecological cycle, with results damaging to the environment. In this case, however, the situation is not a lost case. Something can be done to reduce the pressure by redirecting the nature of the relationship. The region is not a significant producer of greenhouse gases, but its land use activities should spare the more marginal ecosystems, reforest them, and put them on the path to rebuilding their declining texture and genetic pool. For this to be successfully pursued, however, will require that action be taken to create new types of economic activities in the region to absorb a larger ratio of its population in a much smaller geographic space. Industries are such activities. As far as that is concerned, Cameroon's industrial strategy has been less than equitable, with a disproportionate ratio of industries located in the Duala-Edea-Kribi and Yaounde corridor. The country's Vision 2035 strategy for the Western Region on the other hand emphasizes increasing economic growth, jobs, and poverty alleviation through improvements in agriculture and the building of infrastructures. But as this chapter shows, *no amount of improvement in agriculture is going to absorb the region's large population on the area of land available.* Creating industries that can employ half or more of the population will release them from farming and with appropriate advising to eliminate farming in marginal areas. Secondly, making it possible for those that remain in agriculture to access modern agricultural inputs to maximize production would ensure that farm yields are high enough to supply the population with enough food at low prices. That will diminish the need for cultivating any piece of land people can find and will be more welcoming to conservation. Thirdly, developing building plans that require buildings to be built up in multi-storey structures rather than on wide surfaces would save space further, as less land area is needed. This would help solve the sprawl problem and will also make it easier to work on the promotion of greenery between the region's various divisions and districts. The government is currently building affordable multi-storey housing in urban housing projects in the capital cities of each of the country's ten regions that might set an example and perhaps also lead to other conceptions of using space efficiently among urban dwellers. The new multi-storey housing complexes are ecological improvements to past public housing known as SICS in Mendong, Yaounde, or Bonamoussadi, Douala because space is indeed more efficiently used in their construction. Fourthly, the continuing problem with the government's approach to environmental and economic issues is that it is inefficient. It does not properly focus on regions that have environmental problems. It instead looks at all regions, whether they have such problems or not. But the East, Adamawa, the South are places that are not underpopulation, environmental and agriculture pressure. The government's approach, in other words, is more political rather than issue-directed problem-solving. Fifth, the government would need to create a regulatory body to deal with environmental issues rather than leave things in the hands of line Ministries. The Ministry of the Environment should serve such a role, but it is unclear whether it is a regulatory body or a regular line Ministry that ends up dealing with the issues in a political manner.

Household waste disposal constitutes another category of issues that need to be mitigated. For many cities in the region, people dispose of their growing quantities of garbage wherever they can on open dumps. The result is that some of the garbage ends up polluting the areas, and appropriate disposal could remedy that. Waste disposal is the responsibility of municipalities, but that is not a priority for most of them.

Environmental education has been one thing that environmentalists recommend in overcoming various problems in the natural environment. Environmental education would work especially on issues such as sanitary matters, but also in sensitizing and mobilizing schools and localities to engage in actions to transform their localities from dirt surfaces to green lawns. In the Western Region, as in most of Cameroon, the surroundings of people's homes are considered clean if grass covering is cleared out so that only the ground surface remains. Through environmental education, schools should be encouraged to plant their compounds with grass to create lawns. And localities should be encouraged to do same. Such efforts would reduce dust and erosion and when this becomes a practice, it would also generate jobs in the region for lawn care. More importantly, it is quite crucial and advisable to create forests in the region such that each local council area has a number of forested areas that are not just used for recreation but to eventually serve as sanctuaries for some of the animal and other natural species.

A final area where to move forward is more emphasis for the development of clean and renewable energy in the Western Region, such as solar and wind installations. The region as an area of high altitude and mountainous terrain is well positioned. However, to be successful in addressing the region's many environmental challenges requires that greater urgency be accorded it than has been the case over the past half century. To do so requires acknowledging the extent of the problems first, and then to sensitize people why there is a need for change and to come up with concrete plans to change direction in economic activities in agriculture and industry to reduce the ratio of people in agriculture, restore the vegetation, approach human habitation more sustainably, and devise effective institutional mechanisms to manage, regulate, and spur innovation for better adaptation to environmental possibilities and limitations.

Notes

1. These are the areas with significant forest and/or wildlife resources.
2. In fact, as late as 2001, some African countries were accepting waste imports from abroad, while waste removal from cities on the continent was not a concern. See Jennifer Clapp, 'Africa and the international toxic waste trade,' and Tayo Odumosu, 'When refuse dumps become mountains: responses to waste management in metropolitan Lagos,' in Tesi (2001).
3. Cf. Ministry of Forest and Wildlife of Cameroon (2008).
4. Worldmark Encyclopedia of Nations, 2007.
5. These forces include NGOs, international institutions such as the UN and its specialized agencies (UNDP, UNEP) and various global environmental protocols from the Rio to Kyoto to Paris.
6. Until 2008, the administrative regions were known as provinces, and today's Western Region was called 'Western Province.'
7. See OECD/AFDB, *African Economic Outlook—Cameroon*, January 10, 2002.
8. Mphoweh Jude Nzembayie and Futonge Nzembayie kisito (http://Cameroon-tour.com).
9. United Republic of Cameroon Economic Memorandum Report No. 1798CM, volume 2, Statistical Appendix, March 22, 1978.
10. This first came to my attention in the Spring and Summer of 2003 when I was a Senior Fulbright Scholar at the University of Dschang. I then followed up with closer attention in 2008, 2010, 2013, and 2015. In the process of my inquiry, I was also told that in between planting and harvesting, owners of shops or small businesses in the city who also are planters shut up their shops for some hours during the day to remove weeds from their farms.
11. Ibid.

12. Interview Dieudonné Tessie, Agricultural Extension Director, Bafang, July 15, 2015; Author's Observations in the Summer of 2013, Mbouda.
13. Author's observation in 2003, 2013, and 2015; Interview Dieudonné Tessie, Agricultural Extension Service Director, Bafang.
14. Ibid.
15. Ibid.
16. Notes of discussion on a minibus from Mbouda to Bafoussam, July 10, 2015.
17. Ibid.
18. Peoples from other regions of Cameroon who settle in the Western Region are far fewer in numbers than those from the West who reside out of the region. And many from other regions besides the Northwest were brought there by their jobs accordingly (interviews, Summer 2015).
19. The scenery as one passes through the region during pick planting season bears this out.
20. Sirri Alette Ngwa, 51 New Subdivisions Created in Cameroon, *The Entrepreneur*, April 26, 2007. Politics is said to be the reason why the large Noun Division has not been split into two or three divisions. According to many Bamoun elites, I have posed this question to, the Sultan, the traditional ruler of the Bamouns who inhabit the Noun Division is said to be against any such split because of fear that some parts of the division may try to break away or undermine his authority especially if they are in a separate administrative division.
21. Bureau Central des Recensements et des Etudes de Population, Institut National de la Statistique, Cameroun.
22. Table A1, La Republique du Cameroun en 2010, Cameroon Government (Bureau Central des Recensements et des Etude de Popuation).
23. Interview at the Market in Galim, July 13, 2015.
24. See the 1974 Land Law, Land Ordinance Laws 74-1 and 74-2; Also see Focus on Land in Africa (FOLA) Brief: Land Registration in Cameroon (http://www.focusonland.com/fola/en/countries/brief-land-registration-in-cameroon, accessed November 25, 2017).
25. I am thankful to Raymond Mopoho of Dalhousie University for bringing this to my attention.
26. Interview, Mbouda, Cameroon July 12, 2013.
27. Ibid. For an extensive study of the insurgency in Cameroon, see Richard Joseph (1977). Also: Terretta (2014).

28. Table A1, La Population du Cameroun en 2010, Cameroon Government (Bureau Central des Recensements et des Etude de Population).
29. See Barbier (1977). The Nkonjock relocation project in the Littoral Region next to Yabassi seemed to have succeeded as compared to that of the French. This may be due to the forced and disruptive nature of the French scheme.
30. In discussions with various Bamileke outside of the Western Region, very few saw themselves as being never returning to their ancestral homes in Bamileke land, even those who had been away for decades.
31. Mphoweh Jude Nzebayie and Futone Nzemayie Kisito (http://Cameroon-tour.com, accessed October 21, 2017).
32. See Encyclopedia of the Nations—Africa—Cameroon Industry (http://www.nationsencyclopedia.com/Africa/Cameroon-INDUSTRY.html; Nexus Commonwealth Network: Cameroon, http://www.commonwealthofnations.org/sectors-cameroon/business/industry_and_manufacturing/, accessed November 25, 2017).
33. The World Bank, Staff Appraisal Report, Second Western Province Rural Development Project, March 29, 1984.
34. Information for the industries in the Western Region is from Dr. Raymond Mopoho of Dalhousie University in Halifax, Canada (Also see http://cameroon-tour.com/towns&provinces/index.html, accessed June 7, 2017).
35. The National Coordination of the ANCR-NCSA process, National Capacity Self-Assessment in Global Environment Management (ANCR-NCSA Process): *Capacity Building Action Plan of Cameroon in the Implementation of International Environmental Conventions.* Under the supervision of H. E. Pierre Hele.
36. IISD 2004 (http://www.iisd.org); Bele et al. (2011, 377–379).
37. Project Cooperation Agreement (PCA) for the Expedited Enabling Activity (EEA): Cameroon: Preparation of the Second National Communication under UN Framework Convention on Climate Change (UNFCCC) between UNEP and the Government of Cameroon January 2009.
38. Pierre Hele, Minister of the Environment and Nature Protection, Second National Communication on Climate Change to UNFCCC September 2015.
39. Ibid.

40. See International Finance Corporation, Chad—Cameroon Pipeline Project, (http://www.ifc.org/wps/wcm/connect/region__ext_content/regions/sub-saharan+africa/investments/chadcameroon, accessed November 28, 2017); World Bank Report No. AB3842 Project PI09588, Environmental and Social Capacity Building for the Energy Sector, Government of Cameroon Kribi Deep Seaport Project (http://documents.worldbank.org/curated/en/756341468012662478/pdf/Project0Inform1isal0Stage10approved.pdf, accessed November 28, 2017).

Appendix

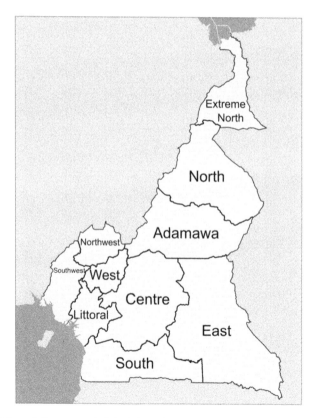

Fig. 7.1 Map of Cameroon's Administrative Regions

References

Barbier, J.-C. (1977). *A Propos de l' Operation Yabassi-Bafang Cameroon.* Yaoundé: ONAREST.

Bele, M. Y., Somorin, O., Sonwa, D. J., Nkem, J. N., & Locatelli, B. (2011). Forests and climate change adaptation policies in Cameroon. *Mitigation and Adaptation Strategies for Global Change, 16,* 369–385.

Champaud, J. (1983). *Villes et Campagnes du Cameroun de l'Ouest.* Paris: ORSTOM.

Champaud, J., & Gendreau, F. (1983). *Migrations et Développement: La Région du Moungo au Cameroun.* Paris: ORSTOM.

Cleaver, K. M., & Schreiber, G. A. (2001). Reversing the spiral: The population, environment, and agriculture nexus in Sub-Saharan Africa. In M. K. Tesi (Ed.), *The Environment and Development in Africa.* Lanham, MD: Lexington Books.

Dongmo, J.-L. (1981). *La Maîtrise de l'Espace Agraire,* vol. 1 of *Le Dynamisme Bamileke.* Yaounde, Cameroun: Centre d'Edition et de Production pour l'Enseignementet la Recherche, University of Yaoundé.

Focus on Land in Africa (FOLA) Brief: Land Registration in Cameroon. http://www.focusonland.com/fola/en/countries/brief-land-registration-in-cameroon. Accessed November 25, 2017.

Global Environment Facility (GEF). (2008). *GEF country portfolio evaluation: Cameroon (1992–2007).* http://www.gefieo.org/sites/default/files/ieo/council-documents/c-34-me-inf-03.pdf. Accessed May 2, 2016.

International Institute for Sustainable Development (IISD). (2004). *Cameroon case study: Analysis of national strategies for sustainable development* (Working Paper, June 2004). Winnipeg: IISD.

Jiotsa, A., Okia, T. M., & Yambene, H. (2015). Cooperative movements in the Western Highlands of Cameroon: Constraints and adaptation strategies'. *Journal of Alpine Research, 103–1.*

Johnson, W. R. (1970). *The Cameroon federation: Political integration in a fragmented society.* Princeton, NJ: Princeton University Press.

Joseph, R. (1977). *Radical nationalism in Cameroon: Social origins of the UPC Rebellion.* Oxford: Oxford University Press.

Kengoum, F., & Tiani, A. M. (2013). *Adaptation and mitigation policies in Cameroon pathways of synergy* (Occasional Paper No. 102). Bogor: Center for International Forestry Research (CIFOR).

LeVine, V. T. (1964). *The Cameroons, from mandate to independence.* Berkeley: University of California Press.

Meadows, D. H. (Ed.). (1972). *The limits to growth.* New York: Universe Books.

Mengang, J. M. (1998). Evolution of natural resources policy in Cameroon. *Yale Forestry and Environmental Studies Bulletin, 102,* 239–248.

Norrington-Davies, G. (2011). *Climate change financing and aid effectiveness Cameroon case study.* Paris: OECD. DRAFT, March 2011. www.oecd.org/dac/environment-development/48458409.pdf. Accessed November 11, 2017.

Rainbow Environment Consult S.a.r.l. (2007). *Inter thematic evaluation of capacity strengthening in the implementation of the three Rio conventions in Cameroon: Biodiversity, climate change, and desertification control* (CSA Project, May 2007). Yaoundé: REC.

Tankou, C. M., de Snoo, G. R., de Iongh, H. H., & Persoon, G. (2013). Soil quality assessment of cropping systems in the Western Highlands of Cameroon. *International Journal of Agricultural Research, 8,* 1–16.

Tchindjang, M., Banga, C. R., Nankam, A., & Makak, J. S. (2005). *Mapping of protected areas evolution in Cameroon from beginning to 2000: Lesson and perspectives.* www.cartesia.org/geodoc/icc2005/pdf/oral/TEMA10/Session%206/TCHINDJANG%20MESMIN%202.pdf. Accessed August 10, 2016.

Terretta, M. (2014). *Nations of outlaws, state of violence: Nationalism, Grassfields tradition, and state building in Cameroon.* Athens: Ohio University Press.

Tesi, M. K. (Ed.). (2001). *The environment and development in Africa.* Lanham, MD, USA: Lexington Books.

Tesi, M. K. (2018). The state, post-independence politics, and the struggle for democracy in Cameroon. In J. Takougang & J. Amin (Eds.), *Post-Colonial Cameroon.* Lanham, MD, USA: Lexington Books.

Tieguhong, J. C., & Betti, J. L. (2008). Forest and protected area management in Cameroon. *Tropical Forest Update, 18*(1), 6–9.

World Commission on the Environment and Development. (1987). *Our common future.* Oxford: Oxford University Press.

Government Publications and Legal Documents

Ministry of Forest and Wildlife of Cameroon (MINFOF). 2008. Programme Sectoriel Forêt-Environnment Rapport Annuel d'activité 2007, une vue globale sur les activités programmées et les principaux résultats attents au courant de l'année 2007, Observations, Limites et Recommendations, Février 2008, Cameroon.

Ordinance no. 74/01 of 6 July 1974, laying down rules governing land tenure in Cameroon / fixant les modalités d'application du régime foncier et domanial au Cameroun.

Project Cooperation Agreement (PCA) for the Expedited Enabling Activity (EEA): Cameroon: Preparation of the Second National Communication under UN Framework Convention on Climate Change (UNFCCC) between UNEP and the Government of Cameroon January 2009.

United Republic of Cameroon Economic Memorandum Report No. 1798CM Volume 2 Statistical Appendix, March 22, 1978.

8

The Impasse of Contemporary Agro-pastoralism in Central Tanzania: Environmental Pressures in the Face of Land Scarcity and Commercial Agricultural Investment

Tadasu Tsuruta

Introduction

Sub-Saharan Africa is rapidly changing from a land-abundant to a land-scarce continent. While struggles for arable land among rural villagers intensified during the last two decades, land grabbing by outsiders has escalated across the region. Land resources, which have been communally controlled in each locality, are increasingly regarded as valuable assets with individual ownership, leading to a plethora of land disputes even in remote villages. In Africa, as elsewhere, most of the recent land problems are closely associated with the commercialization of agriculture (Cotula 2013; Hall et al. 2015). Behind the acute land problems also lies a governmental policy promoting agricultural investment from both within and outside of each country, leading to a rapid 'tractorization', without which large-scale land appropriation would not be feasible.

T. Tsuruta (✉)
Faculty of Agriculture, Kindai University, Nara, Japan
e-mail: ttsuruta@nara.kindai.ac.jp

On the other hand, what makes African land problems unique is the widespread presence of pastoralism in semi-arid and arid areas. As herders' nomadic lifestyles make it difficult to establish land rights, rangelands have always been easily encroached upon by local farmers and outside investors. Wetlands, formerly used as pastures, have been targeted in particular by both villagers and outsiders, especially for commercial crop (and livestock) production (Woodhouse et al. 2000), leading to varying scales of land conflicts. Such conflicts have often been perceived as a confrontation between full-time farmers and specialized pastoralists (Benjaminsen et al. 2009; Mwamfupe 2015). Such a simple farmer–herder dichotomy, which has been prevalent in academic as well as everyday discourses, however, disguises the fact that many societies in semi-arid Africa have an *agro-pastoral* production system in which each household combines livestock keeping and farming (Homewood 2008, 87–91). At the same time, it is becoming increasingly difficult to make a clear demarcation between 'farmers' and 'pastoralists', or to affix either label to a particular ethnic group (e.g. Maasai as 'pure pastoralists'). Ethnographic evidence indicates that former full-time farmers often began to keep a large number of livestock, while increasing numbers of pastoralists embarked upon farming in the process of sedentarization (Toulmin 1983; Little 1987; Babiker 2001; Fratkin 2001; McCabe et al. 2010).

Despite the widespread nature of agro-pastoralism in Africa, this unique production system has attracted insufficient scholarly attention as a key subject for inquiry. The work of Brandström et al. (1979) was one of the first attempts at a systematic analysis of agro-pastoralism as a unique way of livelihood in rural Africa. Focusing on three ethnic groups in Tanzania (Sukuma, Wahi Wanyaturu and Gogo) as typical agro-pastoralists, they tried to identify commonalities and differences in the patterns of the combination of animal husbandry and farming practices of these societies. The central point of their argument is interdependency between these two production systems. While the Gogo showed a limited integration of farming and livestock rearing, the Sukuma and Wahi Wanyaturu tried to increase farm output by using ox-plough or manure and then further accumulated wealth by converting surplus grain into livestock.

While their argument seems to be relevant still and worth reconsideration, the socio-economic situation in which these ethnic groups have found themselves has considerably changed since then. We have to analyse contemporary agro-pastoralism in the context of (1) growing population and intensified competition for land resources, (2) further penetration of a monetary economy into village life, (3) increasing commoditization of livestock and crops, and (4) land appropriation and agricultural investment supported by government policies.

This chapter, then, explores the transformation of agro-pastoralism in the semi-arid Dodoma Region of Central Tanzania, amid an ever-increasing pressure on land resources and an accelerated commercialization of agriculture. To highlight the diverse nature of land-resource problems, two different areas were selected for this study: a densely populated older settlement close to the heartland of the Gogo, the dominant ethnic group (M Village); and a sparsely populated periphery with open frontier areas (Itiso Division) (Fig. 8.1).[1] The former case study focuses on how villagers are coping with challenges stemming from scarcity of land resources, while the latter frontier area case highlights land alienation process by outsiders, both agro-pastoralists and urban investors.

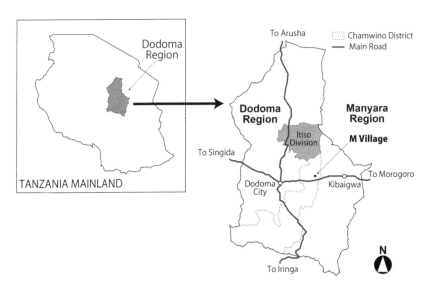

Fig. 8.1 Study area

Background: Historical Overview of Agro-pastoralism in Central Tanzania

Traditional Agro-pastoralism in the Study Area

Dodoma Region has been a marginal area for cultivation, with its annual rainfall rarely exceeding 600 mm, and is often regarded as one of the most 'backward' areas in the country. Chronically beset by crop failure because of low and erratic rainfall, keeping a considerable amount of livestock had been the only available option for the inhabitants to survive a prolonged drought. Before the 1970s, the area was frequently afflicted by famines, which were often minor and localized, although some of them were extensive and devastating (Rigby 1969, 20–22; Maddox 1990). Gogo agro-pastoralists are the dominant ethnic group in Dodoma. There are also a significant number of Maasai and Datoga pastoralists, coexisting side by side with the Gogo, along with agro-pastoral Iraqw who have recently migrated from the north in increasing numbers.

Typical agro-pastoralism can be observed among the Gogo. According to a classical ethnography by Peter Rigby (1969), the Gogo people were 'cultivating pastoralists'. While keeping a considerable number of livestock, they practiced shifting cultivation of sorghum and pearl millet using hand hoes. The field was left fallow after two or three years of cultivation, and the rights to the land lapsed if it was left uncultivated. Although their subsistence depended primarily on agriculture, their value system attached prime importance to cattle. The Gogo's basic residential unit was a homestead composed of several households, each of which was typically made up of a married woman and her children. Despite each household being an independent unit of agricultural production and consumption, livestock was owned by the entire homestead and controlled by its head. Homesteads and relatives were intricately connected through the rights and obligations related to cattle (ibid., 29, 48, 154–187).

Unlike mixed farming developed in Europe, agriculture and pastoralism in the Gogo area virtually had no direct relationship except that the latter provided manure to the former. According to Rigby, there was some evidence that the areas with a greater number of livestock per person tended to have a smaller acreage under cultivation per person. However, livestock could be exchanged for grains in times of drought. Likewise, in times of rich harvest, livestock could be obtained through the sale of surplus crops. In this sense, agriculture and pastoralism had a complementary relationship (ibid., 33, 43–44).

High residential mobility characterized Gogo rural society, being 'an intrinsic aspect of Gogo ecological adjustment and economy' (ibid., 20). The localized nature of droughts was the main reason for the frequent movement of homesteads. Of two neighbouring areas, one may have had a sufficient rainfall for good harvest, while the other may have been hit by a drought. Homesteads moved from one place to another (to areas less affected by drought) in search of pasture and water for their livestock. Hence, there was no clear ownership on the land (ibid., 26–43).

The high mobility of the Gogo people may demonstrate the great flexibility of agro-pastoralism to help accommodate them to a new environment. Another typical agro-pastoralist group in the area, the Iraqw, expanded constantly from their homeland into other areas, and behind this mobile character lies the great ecological adaptability of their livelihood. Fukui (1969, 15) argued that the Iraqw were able to adjust the balance between agriculture and pastoralism flexibly according to ecological conditions. Families living in grasslands put more emphasis on farming than herding, while those in wooded savanna and swamps tended to concentrate more on herding than cultivation (ibid.). Families with different livelihood strategies were connected by an institutionalized network to exchange foodstuff (Loiske 2004).

Post-independence Demographic and Socio-economic Changes in Dodoma Region

The subsistence economy of traditional agro-pastoralism of the Gogo people was transformed dramatically after the 1970s. As elsewhere in Tanzania, an ever-increasing population has put a considerable strain on land resources. The demographic change of the study area is shown in Table 8.1. At the district level, population size and density doubled between 1978 and 2012, with population density reaching 37.3 per sq. km, making it 5.3 times higher than that of 1957 for the entire Gogo land (7 per sq. km, Rigby 1969, 22). In Chilonwa Ward, where M Village was located, the density reached as much as 53 per sq. km, though the growth rate is sluggish recently, possibly indicating that the capacity to accommodate population in this area is approaching its limit. Meanwhile, in another area in focus, namely Itiso Ward, the population grew rapidly after 1988; nevertheless, the density as of 2012 is still relatively low[2] (Table 8.1).

Along with overall population increase, the villagization programme launched by the post-independent socialist government brought about a considerable change in the settlement pattern of the Gogo. 'Operation Dodoma' was started in 1971 to enforce the scattered rural population to live in officially designated villages across Dodoma Region. Such an obligatory resettlement scheme inevitably led to a contradiction with the highly mobile character of the Gogo production system (Mascarenhas 1977; Thiele 1986). In spite of traditional agropastoralism being characterized by frequent relocation of homesteads and extensive use of land, it was becoming increasingly difficult to follow the traditional patterns of land utilization.

A sedentary livelihood pattern brought on by villagization affected the livelihood of Gogo in many ways. Along with the changes in land-use pattern, one of the most fundamental changes was the penetration of a cash economy in every sphere of village life, accompanied by further commoditization of labour, cattle and crops. Thiele (1984, 92) argued that the subsistence economy of the Gogo was transformed from 'natural economy' (where the production of commodities for sale in a market is absent or peripheral) into the 'peasant economy' (where petty commodity production forms an integral part of livelihood), as a result of the villagization programme. Livelihood strategies were diversified (e.g. on-farm wage labour, charcoal burning) according to the economic status of each household as well as the ecological condition of each village. Through this process, the symbolic value of cattle also declined and bridewealth came to be increasingly paid in cash (Thiele 1984, 1986).

Table 8.1 Population increase in the study area

Year		1978	1988	2002	2012
Dodoma Rural District[a]	Population	276,737	353,478	440,565	552,188
	Population density (per km^2)	18.7	23.9	29.7	37.3
	Average annual intercensal population growth rate	–	2.8%	1.8%	2.5%
Chilonwa Ward[b]	Population	8,301	9,100	12,732	14,793
	Population density (per km^2)	30	33	45	53
	Average annual intercensal population growth rate	–	1.0%	2.9%	1.6%
Itiso Ward	Population	4,153	4,953	8,560	12,636
	Population density (per km^2)	8	9	16	23
	Average annual intercensal population growth rate	–	1.9%	5.2%	4.8%

[a]Dodoma Rural District was divided into Chamwino and Bahi Districts in 2006 or 2007. Figures of both districts are combined in the data of 2012
[b]Chilonwa Ward was divided into two different wards somewhere between 1988 and 2002. Figures of both of the new wards are combined in the data of 2002 and 2012
Source Population Censuses, 1978, 1988, 2002, 2012

At the same time, although the region has been hit by drought frequently since the 1970s, famine seems to have been alleviated as villagization has made it easier for villagers to access food aid from the government (Mascarenhas 1977, 378–379).

Commercialization of Agriculture

The pattern of the combination of farming and livestock keeping has changed over time. Formerly, in general, both Gogo and Iraqw homesteads seem to have had an inverse relationship between farm size and the number of cattle in the context of subsistence economy (Rigby 1969, 33, 44; Fukui 1969). However, the commercialization of agriculture, which gradually advanced through the colonial and post-colonial eras, had a great impact on the relationship between these two complementary but discrete production systems. For example, Brandström et al. (1979) found among the Sukuma a 'mutually reinforcing relation between agriculture and pastoralism', in that those wealthy in cattle invested in expanding their farmlands, using extra labour, ox-plough and even a tractor. Similarly, among Iraqw and Gogo, cattle came to be used as a capital to be invested in cash crop production, as I argued elsewhere (Tsuruta 2016). Thus, agriculture and pastoralism, formerly in the relation of inverse proportion, began to have a direct proportional influence on each other in the context of commercial agriculture: the more one accumulates wealth through pastoralism, the more one can expand commercial farming.

Overall development of cash crop production may be, in some measure, responsible for the gradual shift from sorghum and millet to less drought-tolerant maize. As maize became a dominant grain in the domestic market, villagers increasingly preferred it as both cash and subsistence crops. From the 1990s onwards, under the post-socialist market liberalization policy, the maize market, as well as commercial maize production, has continuously expanded owing to private business initiatives, including that for export to neighbouring countries. In the mid-2000s, a modern maize market was constructed in Kibaigwa in eastern Dodoma (Fig. 8.1), as 'a strategic maize marketing center for the country' (Temu et al. 2010). In parallel with the expansion of the maize market, the use of ox-plough gradually increased in the 1990s among the villagers in Dodoma, which is now rapidly being replaced by tractors. Thus, traditional staple grains (sorghum and millet), which were formerly cultivated on a smaller acreage with hand hoes, have mostly been replaced by maize, which is also one of the principal cash crops produced in a larger-scale monoculture.

The rapid expansion of commercial maize production was closely associated with state policies focusing on agricultural modernization and intensification. In particular, the Kilimo Kwanza ('Agriculture First') initiative (2009–) launched by the government to promote agricultural investment made a significant impact on the agricultural landscape in Dodoma. From the Kilimo Kwanza principle, a series of policies were derived, including subsidies for modern agricultural input; tax exemption for imported tractors; and increasing availability of loans for agricultural investment, among others (Mbunda 2011). These policies enlarged opportunities to get farming equipment and access to finance, especially for wealthy investors. In particular, the use of tractors has been encouraged and their numbers increased. The number of imported tractors more than quadrupled between 2008 and 2011 (World Bank 2012, 23).

Changing Agro-pastoralism in a Village Community

Outline of Research Village

This section will examine the transformation of agro-pastoralism in M Village, located some 60 km from Dodoma City (Fig. 8.1), with special reference to the changing pattern of land use and land transactions.

Before the 1960s, residents in the present M Village were widely scattered and divided into several localities. Under the villagization programme, people interspersed in remote mountainous areas were urged to move to a nucleated settlement in the lower part of the area near a mission station by 1973. Since then, the villagers have gradually accepted modern aspects of life such as schools and dispensaries, as well as national language and Christianity. After the villagization policy was officially abandoned in the mid-1980s, a lot of villagers left overpopulated village centres and resettled into their former territories to seek greener pastures and a large acreage of arable land. In this process, remaining woodlands were widely cleared and converted into farmlands. The number of households in M Village increased 2.6 times between 1977 (410 households) and 2009 (1070 households).[3] The major ethnic group in the village is Gogo, living alongside a Nguu minority, who migrated from a distant Tanga Region into this area a generation ago and who accommodated themselves to the existing Gogo culture and lifestyle.

Though agro-pastoralism remains an important livelihood strategy for villagers, several important changes occurred in the agricultural landscape, livestock ownership and land ownership. Along with an increasing population pressure, all these changes are related to the cash economy, which has penetrated into almost all the spheres of village life today. At a glance, the role of livestock as the sole inheritable and exchangeable wealth appears to have been replaced by cash, as can be seen in bridewealth, which is now largely paid in cash. In place of livestock, land is increasingly recognized as the more valuable asset. At the same time, former homesteads, which were units of livestock keeping and were made up of several households, have disintegrated into separate nuclear families.

Changes in livestock keeping and land use are closely related to farming practices which have changed over time in terms of (1) staple crops, (2) use of animal draft power and (3) land-use pattern. As traditional shifting cultivation has become difficult due to the disappearance of fallow land, most villagers, who cannot afford to have an extra field, have begun to use the same farmland consecutively. Though post-harvest grazing of livestock in farmlands may replenish crop nutrients, there is no extensive use of manure in the field. Soil degradation and the ensuing decline of productivity may have been responsible for the expansion of acreage under cultivation, which was made possible by ox-plough.

In M Village, the ox-plough was likely introduced in the 1990s and became widespread in a very short period. Recently, some wealthy villagers with larger landholdings have started to hire tractors. To briefly summarize, former hand-hoe-based shifting cultivation of sorghum and millet has gradually been replaced by maize cultivation on semi-permanent fields based on ox-plough and, later, tractors, adding further pressure on grazing resources.

Agriculture and Food Insecurity

Farming still remains the primary livelihood for villagers, though production is highly unstable as it is easily affected by weather conditions, especially recurrent droughts. Major crops include maize, peanuts, sunflower and sesame. Maize, peanuts and sunflower are grown for both subsistence and commercial purposes, while sesame is a high-value cash crop for export. Other secondary crops include Bambara nuts, a variety of calabash, cucumber, pumpkin, watermelon and cow peas, which are intercropped in maize fields. A variety of leafy herbs grown naturally on the fields is another important group of food repertoire to support the daily diet of the villagers.

Table 8.2 shows the agricultural production of 22 sample households in two hamlets (Hamlets H and S[4]) in M Village, which may be broadly divided into three economic categories: wealthy, middle tier and low tier, corresponding to the number of livestock they own. The average farm size per household for the crop year 2014/15[5] was 7.8 acres, each cultivating two or three separate farm plots. Table 8.2 reveals a striking contrast between production for the year 2013/14 and that of 2014/15. In the year 2014/15, when the area was hit by a severe drought, 13 out of 20 HH which cultivated maize experienced a zero harvest, irrespective of their economic status. Based on his survey in M Village, Kuroda (2016) also reported frequent occurrences of total crop failure of maize during the years between 2006/07 and 2010/11. This may be attributed to a variety of reasons: unstable weather conditions in recent times; a wide application of less drought-tolerant maize; and consecutive use of the same farmland without fallow period owing to the scarcity of land.

Table 8.2 Agricultural production and other income-generating activities of sample farmers in M Village (22 household [HH])

Wealth rank	HH no.	2014/15 Area under cultivation (acre)	Vegetable garden at riversides (acre)	Crop production (debe = 20 litre)[a]						Income-generating activities to cope with food shortage in 2015[b]
				Maize		Sunflower		Sesame		
				2014/15	2013/14	2014/15	2013/14	2014/15	2013/14	
Wealthy	1	24.5	–	0	12	4	0	24	16	L, *mnada* business,[c] sesame
	2	17	–	2	39	0	12	0	6	Motorcycle taxi, tailoring, tobacco
	3	5	Less than 0.25	0	18	–	–	0	6	V
	4	8	Less than 0.25	6	18	0	6	0	6	L, P, V, S, G
	5	8.5	Less than 0.25	2	20	4.5	5	–	–	V
Middle tier	6	12	0.25	0	6	3	12	4	18	L, P, C, F, G
	7	6.5	Less than 0.25	0	18	0	21	0.5	6	L, P, C, V
	8	7	–	0	20	Unspecified	18	2	12	L, P, sale of gravels
	9	3	–	–	12	–	–	0.35	20	L, T, *mnada* business
	10	16	0.25	1	24	1	48	0	18	L, C, motorcycle taxi
	11	5.5	–	0	6	0	–	–	–	S
	12	5	–	0	21	–	–	0	24	G
	13	6	–	0	Unspecified	–	10	0	5	W, C, T, P
	14	5.75	–	7.5	14	12	27	2	–	F, T, P, carrying loads by oxcart
	15	8	–	6	30	7	27	2	12	Tearoom
	16	6	–	0	2	0	3	0	1	Tearoom, G

(continued)

Table 8.2 (continued)

Wealth rank	HH no.	2014/15 Area under cultivation (acre)	Vegetable garden at riversides (acre)	Crop production (debe = 20 litre)[a]						Income-generating activities to cope with food shortage in 2015[b]
				Maize		Sunflower		Sesame		
				2014/15	2013/14	2014/15	2013/14	2014/15	2013/14	
Low tier	17	12	–	1	30	6	9	1	6	P, F, G, peddling tomatoes
	18	1.5	–	0	2	1	2	–	–	G
	19	2	–	0	6	–	–	–	–	G
	20	3	–	0	12	0	–	–	–	G
Uncategorized	21	4.5	0.5	–	–	0.5	108	0.25	14	L, V, C, peddling salt
	22	5	0.5	0	24	1	18	–	–	L, P, V, F, W, C
Average per HH		7.8	–	1.3	16.7	2.5	20.4	2.3	11.3	

[a]As it is difficult to measure area under a particular crop because of the widespread practice of mixed-cropping, only production of each crop is indicated here
[b]Key to sources of income: L (sale of livestock including cattle, goats and pigs); P (sale of poultry, mostly chickens); V (sale of vegetables); C (charcoal burning); F (sale of firewood); T (logging); W (on-farm wage labour); S (brewing sorghum beer or other kinds of alcohol); G (assistance from children, grandchildren or other close relatives)
[c]Mnada business denotes cattle-trading business based on mnada (monthly cattle auction)
Source Field survey, 2015

Though villagers are well aware that the soil in their farmlands has been exhausted, there is no extensive use of cow dung for manure, except for vegetable gardens at the riverside.

Unstable agricultural production naturally leads to constant food shortages. Kuroda (2014, 113) reveals that, in M Village, 25% of sample villagers experienced *njaa* (hunger or food shortage, in Kiswahili) every year between 2002 and 2011, and 75% faced it five times or more. This means that villagers have to rely on outside sources of food and, indeed, the bulk of their food is purchased, except for the years of good harvest. Maize is readily available at local shops, mills or from the stocks of wealthy villagers.[6] To borrow an expression by Kuroda (2016, 175), maize became 'primarily a thing to be bought with money rather than a thing to be harvested from the fields'. In the 2014/15 crop season, two sample households (no. 9 and 21) did not grow maize at all, in anticipation of purchasing maize for cash earned from sesame production (Table 8.2).

There are a variety of ways in which the villagers overcome chronic food shortage. The column at the right side of Table 8.2 shows how sample households were weathering the period of food shortage after a devastating crop failure in the year 2015. Those with livestock and poultry sold them to purchase maize. Those without livestock had to be engaged in various income-generating activities, including the sale of vegetables produced at riversides, on-farm wage labour, charcoal burning, collecting firewood and other small businesses. Charcoal burning is one of the most important sources of income, especially for the poorer sectors of the community, which is putting further stress on the dwindling forest resources in the village, along with brick baking, another important source of income.

Instability of upland farming without a fallow period, scarcity of farmland and needs for cash have driven villagers to shift to hitherto underutilized areas such as swamps, riversides and remaining scrublands on the hills. Seasonal swamps (*mbuga*) and lowlands alongside rivers (*bondeni*) have increasingly been attracting attention as suitable places for agriculture, especially during the dry spells. Fertile riverside lowlands (*bondeni*) are notably used recently for dry-season commercial vegetable production (Table 8.2). Villagers grow maize in the rainy season, producing tomatoes, onions, okra, pumpkins, Chinese cabbage and other leafy vegetables in the dry season, using water from shallow wells. At least three sample households (HH no. 4, 5, 15) who escaped a zero harvest in 2014/15 harvested maize in *bondeni* fields (Table 8.2). A vast tract of grassy shallow wetland (*mbuga*), formerly used as pasture, has been converted to farmlands for maize mono-cropping. In the years hit by droughts, villagers in famine-laden parts of M Village would go to purchase maize harvested in *mbuga* within and outside the village, where one can expect a good harvest, especially in the dry spells, because of the moisture and nutrients preserved in the soil.[7] Meanwhile, the most profitable cash crop, sesame, tends to be grown in newly cleared fields on the hills. Though it contravenes official laws forbidding the clearing of forest or bush on the hills, many villagers venture to reclaim such areas to grow sesame, attracted by a high selling price. However, in the 2013/14 cropping season, to demonstrate the government's strong determination not to allow illegal encroachment on forests, armed officers were deployed to destroy some sesame fields on hillsides around M Village. In the next season, therefore, many villagers had to grow sesame on the flat lands.

Livestock Management: Accumulation of Wealth and Social Differentiation

Given the highly unstable nature of agriculture in the area, livestock remains a major source in the accumulation of wealth, despite the existence of several commercial crops, most notably sesame. However, the number of households keeping or possessing livestock seems to have decreased considerably. While we cannot make a straightforward comparison, in Rigby's study villages, for example, the percentage of homesteads with livestock was as high as 85% in the early 1960s (Rigby 1969, 51). On the other hand, according to my survey in Hamlet H (comprised of 64 households) in 2013, only a quarter (16 HH) raised either cattle, goats or sheep, while the remaining 75% kept neither cattle nor other small stock.[8] Among 10 cattle-keeping households (16%), only six actually owned the animals (Table 8.3). It is noteworthy that

Table 8.3 Livestock keeping in Hamlet H as of August 2013

HH no.[a]	Cattle (Oxen)			Goats and sheep			Pig
	Owned	Entrusted	Total	Owned	Entrusted	Total	
1	45 (8)	–	45 (8)	200	–	200	–
2	–	20 (2)	20 (2)	–	–	0	–
3	–	16 (16)	16 (16)	–	–	0	–
4	–	15 (0)	15 (0)	–	–	0	–
5	–	12 (0)	12 (0)	–	–	0	–
6	5 (2)	–	5 (2)	12	–	12	2
7	4 (4)	–	4 (4)	–	–	0	–
8	2 (2)	–	2 (2)	–	–	0	–
9	2 (2)	–	2 (2)	10	–	10	1
10	2 (2)	–	2 (2)	–	–	0	–
Subtotal for cattle	60 (20)	63 (18)	123 (38)				
11	–	–	–	–	60[b]	0	–
12	–	–	–	25	–	25	3
13	–	–	–	20	–	20	3
14	–	–	–	14	–	14	–
15	–	–	–	10	–	10	–
16	–	–	–	6	–	6	3
Subtotal for small stock				297	60	357	
17	–	–	–	–	–	–	3
18	–	–	–	–	–	–	2
19	–	–	–	–	–	–	1
20	–	–	–	–	–	–	1
21	–	–	–	–	–	–	1
Subtotal for pigs							17

[a]HH (household) numbers here do not correspond to those in Table 8.1, except HH no. 1
[b]The actual owner of these goats and sheep is HH no. 1, who entrusted his animals to HH no. 11
Source Field survey, 2013

37% of cattle and more than 70% of small stock in this hamlet were owned by a single household (HH no. 1). Four households (HH no. 2–5) kept between 10 and 20 cattle, but all of them were not the actual owners of the animals, keeping instead cattle entrusted to them by other persons. The rest (five households, HH no. 6–10) kept a small number of cattle of their own, most of which were oxen. 31% of cattle in this hamlet were oxen, reflecting an increasing importance of oxen as draft animals, especially for ploughing. Nine households (14%) kept traditional small stock (goats or sheep). Pigs were also kept by nine households. Pigs were recently introduced into the village as a lucrative livestock, and nearly half of them were owned by those households without traditional livestock (Table 8.3). Chickens are fairly common nowadays, becoming an important source of income especially for those who do not possess larger livestock.

As is evident from Table 8.3, a glaring disparity can be observed in terms of livestock ownership.[9] The largest cattle owner (HH no. 1), called JM, in his mid-50s, is regarded as one of the richest persons not only in Hamlet H, but also in the entire village. As with most other cattle owners, his fortune was not inherited, but built up single-handedly out of nothing. Being a typical wealthy cattle owner, he has a large family with four wives. Recently, JM has been keeping more than 30 cattle and at least 200 small stock.[10] The herding labour is done by the male members of his extended families including his own brothers, sons and nephews from several separate households. Those who are involved in joint herding are in turn assisted by JM in many ways. JM provides them with (1) foods such as maize and meat, (2) a means of livelihood such as an oxcart that can be hired out, (3) the right to milk the cows, and (4) other necessary financial assistance in solving everyday problems. In a sense, JM's livestock is not his personal, individual property, but is shared by several families related to him, reminding us of a former homestead made up of several different households. Despite the apparent separation of residential units, families are still connected with social bonds which will be demonstrated in times of rituals (weddings, funerals, etc.) and misfortune (crop failure, disease, accidents, etc.). JM is, like a former head of the homestead, responsible for all kinds of familial matters of his extended family.

JM is not just a man rich in cattle, but also a major earner of cash income. He makes an investment in various kinds of business activities, particularly the cattle-trading business at *minada* (local cattle markets, sing. *mnada*). He is a pioneer of *mnada* business in M Village and has encouraged other villagers close to him to be his business partners, who may join the trade using the capital provided by JM. JM also cultivates a larger acreage of land, both to feed his large family and to earn cash from sesame production (HH no. 1 in Table 8.2), although always failing to secure enough food due to unstable weather conditions. JM also plays the role of money lender, mainly loaning cash on standing sesame in the field as collateral. One of his main concerns recently is how to secure a large area for grazing to feed his herd sufficiently. He has then gradually amassed land through purchasing, as we shall see in the next section.

Increasing Transactions of Land

Formerly, customary land rights of the Gogo were very ambiguous. As frequent relocation of homesteads characterized traditional agropastoralism, individual ownership of the land had been neither exclusive nor permanent among the Gogo. The rights to water and land were not inherited. One could retain control of one's farmland as long as it remained under cultivation, and the farmland could be taken by anyone after two years of fallow period (Rigby 1969, 28, 29). In M Village, the former pattern of land use, that was possible only under conditions of low population density and high mobility, gradually changed after the 1990s, when undeveloped scrublands were rapidly disappearing.

Since the Village Land Act (1999) came into force in 2001, all the land in a village is, at least in theory, supposed to be under the control of the village council. According to the official understanding of the M Village government, all the uncultivated land within the confines of the village belongs to the government, which has the authority to take or seize any tract of land remaining uncultivated for more than three years. Land is broadly divided into two topographical categories: 'flatlands' and 'hills'. While flatlands are permitted to be cleared for cultivation, villagers are prohibited from cutting any trees or shrubs on the hills, except for the purpose of collecting firewood and charcoal burning, a lifeline for many poor villagers. If one wishes to cultivate a piece of unused flat land, one has to ask permission of the government to clear it by paying the necessary fees. The ability of the village government to control land, however, is very limited due to the ambiguity of customary land tenure as well as the arbitrary nature of law enforcement by the village authority.

During the last decade, land has become more commoditized and has been increasingly regarded as individual property with exclusive rights. One indication of the commoditization of land can be seen in the increasing cases of leasing land, either for cultivation or grazing. Some villagers (mostly civil servants living in the village) and outsiders (e.g. businessmen from towns) would rent a farmland for either food or cash crop (mostly sesame) production. The normal rent for one season ranges between 10,000 and 15,000 Tanzanian shillings (Tshs)[11] per acre. The land (uncultivated grassland or farmland in dry season) is also leased out as pastureland, in that crop residues (e.g. maize stems) and other grasses on it would be sold at 7000–10,000 Tshs per acre. The rent rose up to as high as 50,000–100,000 Tshs in the dry season of 2015, when a severe drought hit the village, while it dropped sharply in 2016, when grasses were abundant due to unusually heavy rains.

An even more remarkable indication of commoditization of land is in the increasing cases of land purchase. Villagers are normally reluctant to sell their farmland and may sell it only under special circumstances, when they need a large amount of cash urgently. However, trade of land has been increasing since the 2000s, due to the change in circumstances on both sides of supply and demand. The land is normally sold at 10,000–20,000 Tshs (or more according to geographical conditions) per acre (see Table 8.4). It is noteworthy that the selling price of land is not very much different from that of the annual rent as stated above. This may reflect the fact that, as commercial land transactions are only recent phenomena in this village, the distinction between ownership and usufruct is still ambiguous to some extent, leading sometimes to a dispute caused by misunderstandings between two concerned parties.

A handful of wealthy people in the village have recently been accumulating wealth further through land purchases. The aforementioned

Table 8.4 Chronological list of land purchase by JM

Year of acquisition	Case no.	Area (acre)	Price (Tshs)	Price per acre (Tshs)	Land category	Primary purpose of purchase	Estimated reason for sale	Land acquisition process of the original owner
1997	1	4	30,000	7,500	Riverside	Farming	Unspecified	Unspecified
1998	2	0.75	15,000	20,000	Riverside	Farming	Unspecified	Unspecified
2001	3	2	70,000	35,000	Riverside	Farming	Unspecified	Unspecified
2005	4	6	Unspecified	–	Flatland	Farming	Unspecified	Inherited
2007	5	5	40,000	8,000	Riverside	Farming	Payment of fine (?)	Inherited
2007	6	2.5	25,000	10,000	Riverside	Farming	Medical Bill	Inherited
2008	7	20	200,000	10,000	Flatland	Farming	*Falaga* ceremony[b]	Inherited
2009	8	10	90,000	9,000	Flatland	Farming	Debt repayment	Purchased
2012	9	4	80,000	20,000	Flatland	Farming	Unspecified	Purchased
	Subtotal	54.25	550,000	14,938				
2013	10	5	100,000	20,000	Flatland	Farming	Bridewealth	Inherited
	11	2	70,000	35,000	Flatland	Unspecified	Medical bill	Unspecified
	12	2	80,000	40,000	Flatland	Farming	Debt repayment	Inherited
	13[a]	4	80,000	20,000	Flatland	Farming	Investment capital	Purchased
	14	30	400,000	13,333	Hill	Grazing	Debt repayment	Cleared
	15	9	90,000	10,000	Hill	Grazing	Unspecified	Cleared
	16[a]	4	130,000	32,500	Riverside	Farming	Unspecified	Unspecified
	Subtotal	55	950,000	24,405				
2014	17	4	130,000	32,500	Flatland	Sesame production	Debt repayment	Inherited
	18	5	110,000	22,000	Flatland	Sesame production	Medical bill	Inherited
	19[a]	3	80,000	26,667	Hill	Grazing	Food shortage	Cleared
	20	2	20,000	10,000	Hill	Grazing	Government prohibition	Cleared
	Subtotal	14	340,000	22,792				

(continued)

Table 8.4 (continued)

Year of acquisition	Case no.	Area (acre)	Price (Tshs)	Price per acre (Tshs)	Land category	Primary purpose of purchase	Estimated reason for sale	Land acquisition process of the original owner
2015	21	13	200,000	15,385	Flatland	Farming and grazing	Debt repayment	Cleared
	22	5	150,000	30,000	Flatland	Farming and grazing	Debt repayment	Cleared
	23	7	160,000	22,857	Flatland	Grazing and farming	Food shortage	Cleared
	24	3	30,000	10,000	Hill	Grazing	Medical bill	Cleared
	25	3	50,000	16,667	Flatland	Grazing and farming	Medical bill	Cleared
	26[a]	4	58,000	14,500	Hill	Grazing	Food shortage	Cleared
	27[a]	4	50,000	12,500	Hill	Unspecified	Medical bill	Unspecified
	28	2.5	20,000	8,000	Hill	Grazing	Unspecified	Cleared
	29[a]	10	150,000	15,000	Hill	Grazing	Government prohibition	Cleared
	30	15	200,000	13,333	Hill	Grazing	Unspecified	Cleared
	31	3	60,000	20,000	Flatland	Unspecified	Food shortage	Cleared
	Subtotal	69.5	1,128,000	16,204				
	Grand total	193.8	2,968,000	15,319				

[a]These cases indicate that the sellers were close relatives of JM
[b]A memorial service for the deceased
Source Field survey

polygamous family head JM may represent a typical land concentration process by wealthy cattle owners. As shown in Table 8.4, since 1997 he has purchased land whenever the opportunity arises, both for farming and for securing enough grazing lands to accommodate his large herd of cattle and small stock. Table 8.4 reveals that JM's land purchases accelerated particularly between 2013 and 2015, when he was suffering from a serious shortage of pasture. For these three years only (2013–2015), he spent 2,418,000 Tshs, equivalent to the approximate value of only four head of adult cattle, to purchase 139.5 acres, almost 18 times the average area of farmland in M Village. Meanwhile, the table also shows that, before 2013, he was chiefly seeking farmlands by the riverside, where there is a lower risk of crop failure in times of drought (cases no. 1, 2, 3, 5, and 6 in Table 8.4).

In M Village, selling land or the grasses thereon has become one of the quickest ways of getting cash. Normally, those who are in need of cash approach JM, asking him to buy a tract of land in their possession. Table 8.4 shows that there are several common reasons for the sale of the land, to wit, medical expenses, debt repayment, food shortage and preparation for familial ceremonies. Recently, an increasing number of debt-ridden villagers who received a loan from neighbours or a microfinance institution[12] were forced to sell their land to pay back the loan (cases 8, 12, 14, 17, 21, 22). Some villagers sold their land to JM more than once, and at least two of them (HL for cases 14 and 17, and MN for cases 21, 22, 23) were heavily indebted to a microfinance institution. HL borrowed money from JM, using standing sesame as collateral. Neither money nor sesame was paid back to JM, who eventually gained a large acreage of land (30 acres) as compensation (case 14). After the local government cleared sesame fields as a measure to tighten up the prohibition of land use in hilly scrubland in 2014, the sales of land in the 'hill' category increased (cases 19, 20, 24, 26–30).

Purchasing a piece of land, however, does not necessarily secure the permanent ownership for the buyer. Ownership of lands in the 'hill' category may be revoked by the government at any time, though for JM only two or three years of grazing does pay off its cost. Disputes often arise from competing land claims, which may stem from the ambiguous distinction between leasehold and proprietary rights, and between common and individual ownerships. For example, there are some disputed cases in which a villager sold (or leased) the same parcel of land twice to two different individuals, disregarding other members of the family who have rights to the land. By way of precaution, JM sometimes makes a written agreement with the seller to prove the legitimacy of a title to the land, particularly when the seller is not a person to be trusted.

At the same time, the sale of the land does not always mean that one loses the right to use it. JM purchased land from some close relatives who were in need of urgent cash (cases 13, 19, 26, 27, all of them are JM's nephews). In a sense, JM extended a helping hand to them. In the case of 13, a nephew who needed initial capital for his pork business asked his maternal uncle, JM, to purchase his farmland. Even after selling his land, it is possible for him to use this field free of charge, if necessary. Even in the case of a non-relative, JM allowed the original owner to continue cultivating a part of the land without a rent, even after the land changed hands.

Summary

A high degree of instability characterizes farming in M Village, with crop yields fluctuating dramatically from year to year. Even wealthier households barely manage to (and often fail to) secure their subsistence, despite the overall tendency for agricultural commercialization. However, unlike earlier times, famines no longer have such a devastating effect, because food is readily available if one has enough cash to buy it. Ironically, cash is the most important means to get everyday foods in this region, one of the poorest in the country. As off-farm labour opportunities are very limited, livestock still plays an important role as insurance against crop failure, although there is a great disparity in livestock ownership among villagers. There is also a notable expansion of farmlands into seasonal swamps and riversides, leading to a further decline in pastureland.

A handful of wealthy cattle owners have also been consolidating large landholdings through frequent purchases of land (for grazing and cultivation) from those villagers who need immediate cash for medical treatment, purchasing foods, bridewealth, debt repayment, and so forth. Some villagers lost the bulk of their landed property because of indebtedness. Newly introduced microfinance systems require collateral in the form of immovable properties such as land or premises. Thus, some villagers have come to place a higher value on land than on cattle, because the former can be mortgaged to get access to a loan, while the latter cannot. This, however, does not necessarily mean that wealth in cattle will eventually be converted into wealth in land, because the market value of cattle is still much higher than the value of land, due to thriving cattle business.

Commoditization of land and cattle, however, is always incomplete in the context of the village community. In many cases, cattle and land are still shared by several family members, and it is very difficult to have exclusive personal rights to these assets. Those lands indicated as 'inherited' in Table 8.4 required negotiations with other family members prior to the sale. Wealthy cattle owners also have reciprocity obligations, in that they have to redistribute wealth (cattle, cash and land) to close family members and relations, and sometimes to other disadvantaged villagers. The transaction of land was based not only on purely economic logic, as can be seen in the relationship between JM and his nephews. Such moral relations embedded in the village community suggest that villagers are subject to certain social constraints, along with environmental ones, in expanding agricultural production, in contrast to the case of frontier areas in Itiso Division, to which we now turn our attention.

Land Conflicts in a Frontier Area

Outline of Land Problems in Research Area

This section presents an overview of land issues in Itiso Division of Chamwino District, Dodoma Region. Itiso is located on the border between Gogo land and Maasai land (see Fig. 8.1). In the last two

8 The Impasse of Contemporary Agro-pastoralism ...

decades, this hitherto underpopulated area witnessed a great influx of farmers, pastoralists and agro-pastoralists from other regions. These new arrivals may be divided broadly into two categories: (1) large-scale farmers including urban investors who cultivate more than 100 acres (called *wawekezaji*-'investors' in Kiswahili); and (2) smaller migrant families (*wahamiaji*-'migrants'). The former category ('investors') includes farmers hailing from other densely populated areas (especially Southern Highland), and wealthy urban residents such as business people and political elites. The latter category ('migrants') includes Maasai pastoralists and Iraqw agro-pastoralists, both hailing from neighbouring Manyara Region. As shown in Fig. 8.2, which indicates three major flows of migration and investment around Itiso, the movement of people and capital is highly complicated and has created a complex mosaic of different ethnic groups and outside investors, causing a variety of tensions according to the context of each locality.

There are two major reasons behind this rapid increase of immigrants and agricultural investment in Itiso: one is ecological and the other economic. As this area was formerly only sparsely populated, there remained vast tracts of uncultivated land, especially *mbuga* or extensive

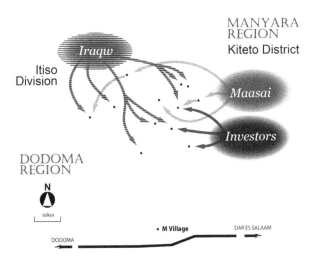

Fig. 8.2 Flows of migration and investment in Itiso

grasslands that stretch along seasonal swamps which were formerly used as pastures. Maasai pastoralists began to settle in Itiso in the early 1980s or earlier, creating separate settlements from local Gogo. From the early to mid-2000s, as the liberalization of the maize trade and penetration of the cash economy advanced, commercial as well as subsistence maize farming expanded in the area by both insiders and outsiders. Kibaigwa international maize market was established in 2004 (Fig. 8.1), becoming a bustling trading centre for maize destined for Kenya and other countries. From around that time, investment of large-scale farmers and immigration of Iraqw agro-pastoralists accelerated. In particular, fertile virgin *mbuga* was targeted by both migrant families and investors, reducing the space available for herders. The greater availability of tractors contributed considerably to the rapid expansion of farmlands. Below, I will give a rough sketch of some features and impacts of two different kinds of land appropriators: 'investors' and 'migrants'.

Expansion of Commercial Farmland by Domestic Investors (*wawekezaji*)

From around 2000, and particularly after the launch of the Kilimo Kwanza initiative, people from other regions have made a considerable investment in commercial maize farming in Itiso Division. According to my extensive survey, investors and farmers hailing from outside were made up of a variety of actors, ranging from influential political elites from both national and local levels, commercial farmers, local business magnates, to lesser businesspersons. These wealthy entrepreneurs lived in Dodoma, Dar es Salaam or other urban centres including Kibaigwa town, employing farm labourers recruited from both local and other areas in Dodoma Region. While original villagers may cultivate within the range of five to twenty acres, these outsiders may grow maize (or other cash crops such as sunflower and sesame) at the range of 200–800 acres or more. Since all the uncultivated land within a village is supposed to belong to the entire village, such a large tract of land was most likely secured through a backstage deal with village executives, which have only the authority to allocate a public land no larger than 50 acres for those in need of land to cultivate.

A similar enclosure of land by outside investors is also observed, albeit on a larger scale, in the neighbouring Kiteto District, Manyara Region (Fig. 8.2), where a series of high-profile land conflicts has arisen mainly between the Maasai pastoralists and commercial maize farmers hailing from outside. The most notable case is that of Emboley Murtangos, a community-based natural resource management area established by several Maasai villages. Large-scale tractor farmers invaded this community reserve and were eventually evicted by force by Kiteto District Council, which was in turn sued by farmers for demolition of private properties. Eventually the District Council, which sided with pastoralists, won in the Court of Appeal in 2011 (Askew et al. 2013). Even after the rule of the Supreme Court, however, some farmers returned there from 2013, and in January 2014, a violent clash occurred between farmers and pastoralists, leading to the deaths of 15 people.[13] To escape from such a bitter confrontation, some large-scale farmers left Kiteto and moved into Itiso.

Even though such a dispute is often represented as an age-old conflict between cultivators and pastoralists, there is, in fact, a more complex relationship between a number of different stakeholders, including small holders and agro-pastoralists, among whom tractor farmers are most responsible for the trouble. One of the issues often overlooked is the fact that agriculture and pastoralism are often conducted by the *same* household, that is, agro-pastoralism.

A Heavy Influx of Migrant Agro-pastoralists (*wahamiaji*)

Along with large-scale land appropriation, Itiso Division has recently seen a growing number of agro-pastoralists and pastoralists migrating from neighbouring Manyara Region. This has resulted in serious land shortage in villages, creating tension between the local Gogo and these migrants, who tend to occupy a larger area of land than the original residents. Notably, Iraqw agro-pastoralists, who have a long history of territorial expansion dating from the colonial era, started to move into Itiso in ever greater numbers from the mid-2000s. As typical agro-pastoralists, they attach a great importance on both farmlands and cattle, which may have aggravated and complicated the land problem in this area.

According to local villagers, a typical process of Iraqw land acquisition is as follows. In the first place, one family arrives in a village, asking villagers to lend or sell them a small plot of land. From the initial small piece of land that they were assigned, they try to expand their territory, encroaching gradually on adjacent areas (for either farming or grazing) without permission from the owners. Later, following the preceding household, more families or relatives arrive, leading to a further encroachment of land. At the beginning, Gogo villagers did not regard land as a valuable property. Iraqw took advantage of this ambiguity of Gogo perception of land tenure, in which it is hard to make a clear distinction between lending and selling.[14] In the eyes of Gogo villagers, Iraqw are possessed by an obsession with land, as they hailed from a homeland with an acute land shortage.

Cattle are the most important mobile property for Iraqw migrants, both as insurance or as cashable assets. They normally come with a small number of cattle and enlarge their herd by using surplus grain. An example of this is as follows. Mr. A left Hanang, the heartland of Iraqw people, with 23 head of cattle and settled in Itiso in 2003. He purchased a small plot of land very cheaply, and proceeded to grow maize and sunflower, income from which he used to buy more cattle as a way of savings. He now possesses 60 head of cattle, in addition to engaging in cash crop production.

Some wealthy Iraqw use tractors to cultivate hundreds of acres, principally growing maize. Gogo villagers also insist that it was Iraqw migrants who brought ox-plough (and then tractor) cultivation to *mbuga*, which was formerly never used as farmland. Generally, Iraqw villagers now possess more cattle and a larger area of land than the original (mainly Gogo) inhabitants. A factor that makes the land problem in this area complicated and difficult to solve is the fact that these migrants are agro-pastoralists, who need a vast acreage of land for both farming and grazing. As a local officer in charge of natural resources puts it:

> In Itiso, the number of livestock keeps on increasing, and there is no reason or sign for decreasing. Livestock is like a bank in which one can save and accumulate wealth. If one sells one head of cattle at 500,000 Tshs, he may buy two calves at 150,000 Tshs each and the remaining

cash (200,000) will be used for other expenses. The same is applied to small stock like goats. And nowadays you have cash crops which are more expensive than food…those who keep a large herd of cattle are equivalent to those who have enough capital or equipment to operate a large farm. When one gets surplus from cash crops, then he further multiplies his herds. Formerly, once a famine broke out, the number of livestock considerably declined. But today it is not the case, because food is abundant everywhere.

Some villages in Itiso were demarcated to cater to Maasai pastoralists as grazing areas. However, as maize has now become an indispensable part of their diet, they adopted cultivation, notwithstanding the typical image of Maasai as genuine pastoralists. Recently, some wealthy Maasai purchased or hired tractors and grew cash crops in a large acreage, further evidence of mutually reinforcing relationships between farming and pastoralism. Due to the expansion of farmlands, some Maasai lost enough pastureland and had to invade forest reserves in Itiso Ward, causing further deforestation.

Conclusion

Agro-pastoralism is one of the most prevalent means of livelihood among people in the semi-arid regions of sub-Saharan Africa. In the focus area of this chapter, Dodoma Region of central Tanzania, Gogo agro-pastoralists have confronted several challenges which make it difficult to maintain the semi-nomadic way of 'traditional' agro-pastoralism. Nowadays, the bulk of the rural population resides permanently in densely populated nucleated villages, and many villagers have ceased to keep livestock, having come to depend on agriculture, which is highly unstable in a semi-arid climate. As food is always available in the market these days, villagers have increasingly engaged in income-generating measures other than farming and livestock keeping, many of which are dependent upon dwindling forest resources. Meanwhile, land is now concentrated in the hands of a small number of wealthy cattle owners, due to a growing scarcity of grazing land.

While 'traditional' agro-pastoralism is becoming infeasible for ordinary villagers on the one hand, a 'commercialized form' of agro-pastoralism only aggravates the land problems even further. A shortage of rangeland is further exacerbated by rapid expansion of commercial maize farms into seasonal swamps, which were formerly used mainly for grazing. The expansion of maize monoculture is partly enhanced by the agricultural modernization policy of the Tanzanian government which allows easier access to imported tractors and agricultural loans. Along with large-scale investors, wealthy agro-pastoralists (including Maasai) also require vast tracts of land for both mechanized farming and grazing to cater to a large herd. Thus, the more one accumulates wealth through pastoralism, the more one can expand commercial agricultural production. There can be no doubt that this chain reaction creates considerable pressure on land use in the area.

Expansion of farmlands, as well as intensive grazing of livestock, has put a considerable amount of pressure on the environment. The loss of vegetation cover easily leads not only to active soil erosion, but also to flash floods in instances of heavy downpours, resulting in recurrent gully erosion which can be observed in both M Village and Itiso. Rapid deforestation in the upstream of rivers may be partly responsible for the increasing cases of serious floods in the downstream. An example of this can be seen in Kilosa District of the neighbouring Morogoro Region, located roughly 100 km downstream of M Village (and 150 km from Itiso), which experienced severe floods in 2009 and 2016, causing a wide displacement of its population.

Most formulas of rural development are based on the experiences of societies with agricultural specialization which cannot be applied directly to rural Africa. A policy like Kilimo Kwanza, focusing solely on agricultural intensification, dismisses the fact that rural livelihoods in Africa encompass many non-agrarian elements, including pastoralism, hunting and gathering, and other diversified subsistence as well as income-generating activities. This kind of policy also ignores the fact that farming practices and landholding in African villages have a number of features different from other agrarian societies (such as modern Europe) characterized by mono-cropping, mixed farming, higher use of input and individual ownership of farmland. We have to seek, therefore, diversified ways of rural development better based on local ecological conditions and existing skills, while taking heterogeneous livelihood and land-use patterns into consideration.

Notes

1. The data for M Village were collected through 80 days of fieldwork carried out by the author at irregular intervals between 2012 and 2016. In Itiso Division, several preliminary field surveys were made in 2015 and 2016 at several selected villages, where short interviews were made with village authorities and ordinary villagers in order to understand the geographical differences of land problems in this area.
2. The low population density in Itiso Ward may be attributed in part to the existence of a large tract of forest reserves.
3. The figures are given by my co-researcher Mr. Makoto Kuroda. See also Kuroda (2016).
4. While all the sample households in Hamlet H are the Gogo, those in Hamlet S are the Nguu ethnic minority. These two ethnic groups are, however, indistinguishable in terms of livelihood.
5. Cropping season in the area normally begins in November or December and lasts until May or June.
6. From 2015, local shops in the village started to sell milled flour of maize, which had hitherto been only available in towns.
7. Conversely, a *mbuga* farmland has the high risk of flood in years with abundant rainfall.
8. The proportion of livestock-keeping households is higher in some other hamlets on the peripheries of the village, which have been settled relatively recently (Kuroda 2016, 157).
9. Hasegawa (2002) also observed an increasing disparity among homesteads in terms of cattle ownership in Bahi area, west of Dodoma City, in the early 2000s.
10. The number of livestock fluctuates season to season, owing to sales, purchase and change of trusteeship contract.
11. One US dollar was equal to Tshs 1660 as of August 2014.
12. Two villagers groups for a microfinance institution exited in 2015, which were organized in 2011 for the first time in M Village. In the case of default, the unpaid loan is recovered through the sale of the mortgaged property.

13. Mussa Juma, 'Kiteto Victims Ask for Food', *The Citizen*, 22 January 2014.
14. In some cases, original buyers (Iraqw) of land have left Itiso for other areas, leaving the piece of land for remaining families to cultivate, who in turn find themselves in disputes with the children of the original seller (Gogo) of the land, who may be deceased. Ambiguity of land ownership is responsible for this kind of confusion.

References

Askew, K., Maganga, F., & Odgaard, R. (2013). Of land and legitimacy: A tale of two lawsuits. *Africa, 83*(1), 120–141.
Babiker, M. (2001). Resource competition and conflict: Herder/farmer or pastoralism/agriculture? In M. Salih, et al. (Eds.), *African pastoralism: Conflict, institutions and government*. London: Pluto Press.
Benjaminsen, T. A., Maganga, F. P., & Abdallah, J. M. (2009). The Kilosa killings: Political ecology of a farmer-herder conflict in Tanzania. *Development and Change, 40*(3), 423–445.
Brandström, P., Hultin, J., & Lindström, J. (1979). *Aspects of agro-pastoralism in East Africa* (Research Report No. 51). Uppsala: Scandinavian Institute of African Studies.
Cotula, L. (2013). *The great African land grab? Agricultural investments and the global food system*. London: Zed Books.
Fratkin, E. (2001). East African pastoralism in transition: Maasai, Boran, and Rendille cases. *African Studies Review, 44*(3), 1–25.
Fukui, K. (1969). Ecological observation of agro-pastoral people: On the migration and settlement of the Iraqw in Tanzania. *Journal of African Studies, 9*, 1–18 (in Japanese).
Hall, R., Scoones, I., & Tsikata, D. (Eds.). (2015). *Africa's land rush: Rural livelihoods and agrarian change*. Oxford: James Currey.
Hasegawa, T. (2002). *Contemporary development of agro-pastoralism and its sustainability: A case of the Gogo, Tanzania* (MA thesis submitted to Graduate School of Asia and Africa Area Studies, Kyoto University [in Japanese]).
Homewood, K. (2008). *Ecology of African pastoralist societies*. Oxford: James Currey.
Kuroda, M. (2014). *Nzala Pesa*: A current concept of Njaa among the Gogo of Central Tanzania. In K. Sugimura (Ed.), *Rural development and moral*

economy in globalizing Africa: From comparative perspectives (*Proceedings of 6th International Conference on African Moral Economy*, 26–28 August 2013, University of Dodoma, Tanzania) (pp. 105–122). Fukui: Prefectural University.

Kuroda, M. (2016). The dietary patterns and livelihood strategies of rural households in semi-arid Tanzania: Some findings from a household diary survey in Majeleko. In S. Maghimbi, K. Sugimura, & D. G. Mwamfupe (Eds.), *Endogenous development, moral economy and globalization in agro-pastoral communities in Central Tanzania* (pp. 152–177). Dar es Salaam: Dar es Salaam University Press.

Little, P. D. (1987). Land use conflicts in the agricultural/pastoral borderlands: The case of Kenya. In P. D. Little, M. M. Horowitz, & A. E. Nyerges (Eds.), *Land at risk in the third world: Local-level perspectives*. Boulder, CO and London: Westview Press.

Loiske, V. (2004). Institutionalized exchange as a driving force in intensive agriculture. In M. Widgren & J. E. G. Sutton (Eds.), *Islands of intensive agriculture in Eastern Africa* (pp. 105–113). Oxford: James Currey.

Maddox, G. (1990). *Mtunya*: Famine in Central Tanzania, 1917–1920. *Journal of African History, 31*(2), 181–198.

Mascarenhas, A. C. (1977). Resettlement and desertification: The Wagogo of Dodoma District, Tanzania. *Economic Geography, 53,* 376–380.

Mbunda, R. (2011). *Kilimo Kwanza and small scale producers: An opportunity or a curse?* (A Research Report). Dar es Salaam: HAKI ARDHI.

McCabe, J. T., Leslie, P. W., & Deluca, L. (2010). Adopting cultivation to remain pastoralists: The diversification of Maasai livelihood in Northern Tanzania. *Human Ecology, 28,* 321–334.

Mwamfupe, D. G. (2015). Persistence of farmer-herder conflicts in Tanzania. *International Journal of Scientific and Research Publications, 5*(2), 1–8.

Rigby, P. (1969). *Cattle and Kinship among the Gogo: A Semi-pastoral society of Central Tanzania*. Ithaca: Cornell University Press.

Temu, A. E., Manyama, A., & Temu, A. A. (2010). Maize trade and marketing policy interventions in Tanzania. In A. Sarris & J. Morrison (Eds.), *Food security in Africa*. Rome: FAO.

Thiele, G. (1984). State intervention and commodity production in Ugogo: A historical perspective. *Africa, 54*(3), 92–107.

Thiele, G. (1986). The Tanzanian villagisation programme: Its impact on household production in Dodoma. *Canadian Journal of African Studies, 20*(2), 243–258.

Toulmin, C. (1983). *Herders and farmers or farmer-herders and herder-farmers?* London: Overseas Development Institute (Pastoral Network Paper 15d).

Tsuruta, T. (2016). Agriculture-pastoralism complex in historical perspective: A case of Northern and Central Tanzania. In S. Maghimbi, K. Sugimura, & D. G. Mwamfupe (Eds.), *Endogenous development, moral economy and globalization in agro-pastoral communities in Central Tanzania* (pp. 37–58). Dar es Salaam: Dar es Salaam University Press.

Woodhouse, P., Bernstein, H., & Hulme, D. (Eds.). (2000). *African enclosures? The social dynamics of wetlands in Drylands*. Oxford: James Currey.

World Bank. (2012). *Agribusiness indicators: Tanzania*. Washington, DC: The World Bank.

9

Down by the Riverside: Cyclone-Driven Floods and the Expansion of Swidden Agriculture in South-western Madagascar

Jorge C. Llopis

Introduction

Historically, various forest agrarian systems have been and still are present across Madagascar. Examples are *tavy* and *jinja* for the cultivation of rice and other subsistence crops in the central highlands and the northern and eastern areas of the island, and *hatsake* in the southern and western regions of the country, most commonly employed for growing maize. These agricultural systems have been used during generations as a complement to other rural livelihoods. As in many cases they involve the clearance of old-grown forest in the first stage of the cultivation cycle, these forms of agriculture are often considered the major cause of deforestation in Madagascar (Waeber et al. 2015). Their prevalence contrasts with trends witnessed in other world regions where such practices have until recent decades played a major role in the forest-agriculture frontier dynamics (van Vliet et al. 2012). A wealth of scholars has in recent years

J. C. Llopis (✉)
Centre for Development and Environment, University of Bern,
Bern, Switzerland
e-mail: jorge.llopis@cde.unibe.ch

analysed the complexity of the underpinnings bolstering these agricultural practices, for example in the southern and western regions of the island (e.g. Brinkmann et al. 2014; Casse et al. 2004; Réau 2002). However, despite the number of studies evidencing the intricacy of factors sustaining these livelihood strategies currently, simplistic explanations for their persistence in the face of an increasing number of developmental and nature conservation interventions aiming at curb their expansion are still widely accepted. As explained elsewhere (Scales 2011), particularly prevalent has been the view that the very existence of these practices nowadays is due to an unchecked population growth pushing impoverished rural inhabitants onto decimating their natural resources base as a way to survive.

Indeed, the population growth rates and the perceived environmental decline in the country in the last decades have paved the way for these neo-Malthusian narratives to enjoy a privileged position when explaining environmental change dynamics on the island, a perspective held up to the present-day by many international conservation and financial institutions. For example, such a view already permeated Madagascar's National Environmental Action Plan (NEAP), a comprehensive although ambiguous policy guidelines package intended to address the environmental and socio-economic problems of the country all at once, and one of the first such initiatives to be undertaken in Africa (Mercier 2006). The NEAP was launched in 1991 largely supported by the World Bank, with the concrete goal of '…. halting the degradation spiral by reconciling population with its environment' (World Bank 1988, R-2[1]). This view has since pervaded through many of the documents released by the Bank about the socio-economic and environmental situation of Madagascar (e.g. World Bank 1996, 2003), altogether an institution already exerting increasing influence on the country's development and nature conservation policies since the mid-1980s (Kull 2014; Waeber et al. 2016). Despite the Bank gradually broadening the range of factors in consideration when assessing socio-environmental dynamics on the island, the narrative equating the tandem poverty-population growth to environmental degradation still resonates strongly in its recently published 'systematic country diagnostic' (World Bank 2015). In the document, it can be read: 'Poverty in Madagascar is closely linked to environmental degradation. ... Degrading lands are often a consequence of poverty-driven short-sighted land management decisions, including deforestation and over-tilling of soils'. (ibid., 79), or '.... the high rates of deforestation ... are ... an effect of poverty driven agricultural extension practices ... re-enforced by the widespread application of customary law in forest management, which allows for slash and burn as a mean to sustain livelihoods as population grows ....' (ibid., 87).

However—and without denying the problematic role that population growth plays in the environmental dynamics of the country—it can be particularly obscuring to focus on an aggregated understanding of poverty as a direct driver for the extension of some of the agricultural practices considered to be responsible for environmental degradation on the island. Furthermore, and as in the above-mentioned diagnostic, running the risk of hindering the comprehension of the diverse nature of the various forest agrarian systems existing throughout the island, these farmers' practices are all too often lumped together under one-fits-all concepts. This is the case with 'slash-and-burn agriculture', a term often employed to define all these forms of agriculture in the country, but which uncritical use might prevent us from grasping the multitude of environmental, socio-economic and political factors shaping the context within which land use decisions by rural farmers are made. Besides the emotional component embedded in the notion, the inaccurate resort to the 'slash-and-burn' term contributes to conflate just one part of the agricultural process, namely clearance technique, with the whole agricultural system (Scales 2014; Ruthenberg 1971). More importantly, this simplification may result into deflecting our attention from understanding the social and economic root causes of environmental change, a critical knowledge particularly needed when devising development and conservation interventions that are intended to strike a balance between supporting sustainable rural livelihoods and maintaining the island's biological wealth. Indeed, in many cases in Madagascar, these agriculture practices would be more usefully defined as 'shifting cultivation' systems, as the term captures a wider range of the socio-economic and ecological determinants affecting the decision-making processes of rural land users. For agrarian systems where the shifting character of the spatial dynamics is non-existent or difficult to determine and fire is employed as a method of vegetation clearance, the more nuanced term 'swidden agriculture' has been proposed (Mertz et al. 2009) and will be accordingly used through the following pages.

On the basis of recent Madagascar field data, this chapter will situate the neo-Malthusian hypothesis within a broader array of factors that, besides population growth, are perceived by local farmers themselves as relevant pertaining the land use decisions they take. The concrete objective is to investigate the potential of such factors to help in explaining the expansion of swidden agriculture in south-west Madagascar, with a particular emphasis on exploring the influence of aspects seldom addressed in the literature on which conservation and development interventions in the region are based.

Cyclones and Forest. And People

In spite of Madagascar being one of the countries in the planet receiving more cyclones (Hochrainer-Stigler et al. 2015), these extreme weather events remain underexplored (Waeber et al. 2015), especially in regard to their effect on land use management decisions by populations impacted by such phenomena. For instance, while the World Bank diagnostic (2015, 89–91) recognizes the effect of these natural hazards upon the livelihoods of populations in the country, it falls short of analysing the concrete local responses triggered by such events.

Within the scientific literature, most works studying these meteorological phenomena in Madagascar have focused on their effects upon forest dynamics (Lewis and Bannar-Martin 2012; Ganzhorn 1995; Burivalova et al. 2015) and on how these processes in turn affect the fauna to which these ecosystems are habitat (Rasamimanana et al. 2000; Ratsimbazafy et al. 2002). Fewer studies have addressed the effect of cyclones on vulnerable human populations (Brown 2009; Harvey et al. 2014), while research has scarcely investigated the specific feedback relationship between the incidence of these events on rural communities relying on forest-based livelihoods and environmental change. To date, this issue has been only touched upon in the wetter regions of the north-eastern coast, e.g. by Urech et al. (2015) and Brimont et al. (2015), and by Minten et al. (2006) in the central highlands. However, this issue has never been addressed in the drier South-west, the region by far losing more forest in the last decades (ONE et al. 2013), and one of the few where agriculture fields continue expanding into the forest. The following pages will try to shed light on the nexus between cyclone-driven floods and the expansion of forest agrarian systems in the south-west region of Madagascar to illuminate its implications for some of the environmental change processes witnessed there in the last decades.

Fieldwork in Behompy

This chapter presents part of the research conducted between October 2013 and March 2014 in the rural commune of Behompy in south-western Madagascar. Communities here live by one of the largest tracts of spiny forest remaining in the region, an area that since the beginning of the 1990s has been experiencing an intense and increasing process of forest retreat. Despite a protection scheme implemented recently in a joint initiative by the Malagasy administration and the World Wide Fund for Nature (WWF), these forests are still witnessing a trend of degradation and decline, which is believed to be primarily due to the expansion of swidden agriculture and charcoal production. At the time that fieldwork was conducted, local populations were still recovering from cyclone Haruna, which hit the area in February 2013. Exploration of the consequences that this event had on communities in Behompy served as an entry point to historically trace back the influence of cyclones upon the spread of forest-based livelihoods in the area. Besides these climatic events, in recent years populations in Behompy have been also submitted to the recurrent impact of droughts and locust invasions, and most critically to the rampant incidence of rural insecurity. Although during the fieldwork a quantitative socio-economic survey was also completed in the commune, the present chapter is mostly based on the qualitative data obtained thanks to the nearly forty semi-structured interviews and focus group discussions carried out in the region. Whereas the bulk of interviews was conducted with local authorities, farmers and cattle herders, also regional forest and rural development agencies' representatives, and international conservation practitioners working in the area contributed as respondents to the research. With the aim of triangulate findings from the fieldwork, secondary data were obtained from several institutions in the regional capital, Toliara, and in the country's capital, Antananarivo.

Rural Livelihoods Under Pressure

Located 25 km inland from the coastal city of Toliara[2] and upstream the Fiherena river, the rural commune of Behompy is home to some nine thousand inhabitants spread over ten *fokontany*[3] on both sides of the river. The mixed subsistence-market economy in the commune is largely based on agriculture, with two main systems dominating the socio-economic landscape. On the one hand, permanent agriculture is practised on the *baiboho*, alluvial fields located on the river banks and fertilized once a year by the rise of the water during the rainy season, roughly from October to March. The range of subsistence crops planted in the *baiboho* includes several pulses, cucumbers and sweet potatoes, while as cash crops are cultivated sugar cane, cotton and a variety of spices, mostly sold to the *Karana* community in the region (the Indo-Pakistani community, established in Madagascar since the seventeenth century). Maize, which is also cultivated on the *baiboho*, plays a double role as a staple crop and as an object of trade. For several crops including maize, two harvests a year could be obtained until a few decades ago.[4] However, nowadays water scarcity constrains cultivation on the *baiboho*, making difficult to obtain more than one harvest per year for most of these crops, concretely because of the erratic character of rainfall and that no permanent irrigation scheme is available in the commune.[5]

The other agricultural system in the commune is *hatsake*, swidden agriculture carried out on the limestone uplands surrounding the river, whereby people resort to slash-and-burn to clear the vegetation and fertilize the soil. *Hatsake* is primarily employed to grow rain-fed maize, of which a significant portion is marketed either locally or most commonly in the nearby city of Toliara. The *hatsake* cycle begins by clearing old-grown forest with a small axe during the dry season, the trunks left to dry and then set on fire before the first rains arrive, a procedure through which the nutrients contained in the ashes are transferred to the soil. This technique also prevents weeds to appear in the first years of cultivation as their seeds are destroyed during the combustion of the trunks while allowing for fertilization of the rather unproductive limestone soils. Before the first rains of the season or just after them, maize is sown on these plots, which during its growing period requires almost no labour, unless acute lack of rain demands some watering. After three or four years of continued maize cultivation, these plots are either (most commonly) left fallow, planted with less demanding crops such as groundnuts or cassava for a few more years, or more rarely converted into semi-permanent agriculture plots in case the biophysical configuration of the plots and water availability are favourable. Given the relatively recent origin of the widespread practice of *hatsake* in the commune, and the fact that it has been calculated that under current conditions biomass regrowth can take more than twenty years to permit another productive burning-cultivation cycle (Raharimalala et al. 2010), the actual shifting nature of swidden agriculture as it is nowadays carried out in Behompy (e.g. whether the plots left in fallow are used again after vegetation regrowth) is difficult to determine. For this reason, the regional noun for this practice, '*hatsake*', or 'swidden agriculture', is the term used throughout this chapter.

An inexpensive and labour-saving system with relatively high yields during the first years of cultivation, *hatsake* has been traditionally employed to complement the other livelihood activities of rural inhabitants throughout southern and western Madagascar (Fauroux 2000). However, at the present time, given the pace at which *hatsake* expands in the commune and the biophysical and climatic features prevailing on the limestone plateau, its spread poses a challenge to the continuity and regrowth of the spiny forest, which tends to revert to an open grassland landscape (Grouzis and Milleville 2001). The fast depletion of the nutrient-poor limestone soils compounded in parallel by weed invasion drives a sharp drop in yields after the third year of cultivation, encouraging farmers to abandon the plot and clear a new field in the forest to keep sufficient returns to their labour. Moreover, despite that *hatsake* has been found to yield up to two tons of maize per hectare in wetter areas of the region (Réau 2002; Milleville et al. 2001), farmers in Behompy reported yields well below this. The main reason given is the scarce and unpredictable character of rainfall, rendering cultivation of maize under a rain-fed regime in this area highly susceptible to harvest failure. It is estimated that maize needs a minimum of 500 mm of rainfall during its growing period to properly develop (Scales 2011; FAO 2013), a condition only met in seven occasions in the area in the last forty years,[6] while the mean annual precipitation was around 300 mm.[7] As perceived by inhabitants nearby this area (Gardner et al. 2015) and also in Behompy, rainfall is declining and has become more unpredictable in the past decades. This perception fits in with the trend observed consistently throughout south-western Madagascar at least since the early 1980s, with drought episodes, once deemed to have a cyclical nature, considered to become chronic (Casse et al. 2004; Minten et al. 2006).

Besides agriculture, zebu rearing is the second most important economic activity in the commune and a key cultural foundation of the Masikoro people, the ethnicity of most inhabitants in Behompy, who define themselves as agropastoralists (Tahirindraza 2006; Razafimanantsoa 1987). Alongside their use as draught power animals for transport and ploughing, and their economic role as a source of savings to be expanded in prosperous periods and to be resorted to in times of hardship, cattle have a prominent place in the cultural life of the commune, as in southern Madagascar in general (Réau 2002; Heland and Folke 2014). Zebu is transacted or sacrificed on several occasions in the life of a Masikoro individual, from the circumcision rite performed in adolescence to being a central element of the dowry in marriages up to the funeral, when a zebu would be slaughtered and partaken of in a common meal by the community. Given their multiple roles, cattle are a prominent sign of wealth and a precious object of status, and as such also a key target for thieves. A well-known phenomenon in southern Madagascar, cattle rustling has traditionally played an articulating role in land tenure arrangements and even is a 'rite of passage' among several of the ethnic groups in the region (Saint-Sauveur 1996; Hoerner 1982; Fauroux 1989). However, in the last few years, the incidence of violent rural banditry attacks involving cattle thefts has acutely intensified, both in the number of assaults and in their death tolls (Pellerin 2014).

Next to with agriculture and zebu husbandry, the third major economic activity and livelihood pillar of the inhabitants of Behompy is the production of charcoal, which is together with firewood the main source of domestic energy for the overwhelming majority of households (Montagne et al. 2010). Apart from a small share that producers keep for their own consumption, charcoal obtained from the forests is either sold in the villages nearby or shipped out to meet urban demand in Toliara. This activity possibly constitutes the second most important source of income for households in Behompy after agriculture, given that more households were engaged in charcoal production than in cattle raising, which is also decreasing in commercial significance (Fauroux 1989; Rasambainarivo and Ranaivoarivelo 2006). Despite that the research conducted in the commune did not capture the temporal dynamics of charcoal production within the livelihoods portfolio of households, in the same area a sharp increase of this activity has been recently reported (Gardner et al. 2015), with predictable implications for the forest cover. This increase was correlated to a decrease in cultivation and pastoralism as sources of revenue, while the specific factors found to be responsible for this trend were the increasing scarcity of rainfall in the last years and the rising incidence of rural insecurity, as perceived by local populations.

The Fiherena River

Stretching over 200 km from the Isalo Mountains on the border with the central highlands up to the Mozambique Channel, the Fiherena river is known for its ferocity, particularly after heavy rains, causing severe damage throughout its flash flood-prone watershed (Chaperon et al. 1993; Olson 1984). Even in the case of cyclones occurring or passing through the southern highlands or nearby areas, the flooding effect of these climatic extremes can be felt already after a few hours of heavy rainfall upstream, particularly on the alluvial plains on the river's lower course. Roughly 40 km of the river runs across the boundaries of the commune of Behompy, where along its banks are located over 600 ha of the *baiboho* fields, on which the local populations most rely to earn a living.

The history of the communities living in Behompy is shaped by the Fiherena and its unpredictable behaviour. In the early twentieth century, one of these flooding episodes led to the prohibition of rice cultivation in the commune,[8] crop which is the main staple in Madagascar. At that time, the river inundated the nearby villages during several days and it was said that it would not be until the communities performed a ceremony under the guidance of the *ombiasy* (diviner or traditional healer), who, among other measures, declared the prohibition of rice cultivation, that the river came back to its normal course. Years later, the road constructed in the 1920s by the colonial authorities to connect the city of Toliara with the highlands, originally laid down along the riverside and through Behompy, had to be moved to its current location, across the Belomotse and Vineta plateau, as the effects of the floods meant an unbearable burden for the road maintenance (Tahirindraza 2006; Hoerner 1979).

More recently, in December 1978 Cyclone Angèle caused the Fiherena to flood the commune of Behompy, destroying hundreds of hectares of agricultural land up to the city of Toliara, where it left dozens of casualties (Hoerner 1979; Olson 1984). Angèle also destroyed the section of the irrigation scheme located in Behompy, dating from the 1950s. It was an extension of the scheme constructed under the colonial regime since the 1920s and beginning in the plains in the downstream river area[9] (Tahirindraza 2006). Despite that projects were undertaken in the last decades to expand or rehabilitate irrigation schemes in the area, particularly after damages by cyclones (World Bank 1995; AfDB 2015), this has not been the case with the irrigation section in Behompy, which has remained unrepaired up to the present. As in other areas of the region, the deterioration of irrigation infrastructure might have had important implications for forest cover (Gardner et al. 2015; Virah-Sawmy et al. 2014), the decline of which could have become more acute after the disengagement of the Malagasy state from the maintenance and rehabilitation of these schemes from the 1980s onwards. Concretely, this disengagement would have been a result of first, the conditionalities attached to the loan packages provided by the International Monetary Fund and the World Bank throughout those decades, and second, by the wave of economic liberalization and state decentralization policies implemented in the following years (Tahirindraza 2006; Hoerner 1990; Minten et al. 2006; Marcus 2007).

The 1980s and 1990s: *Baiboho* Silting, Drought, and a Maize Boom

Cyclone Angèle also constituted a tipping point for the agro-environmental dynamics in the commune. Overlapping with the deforestation trend in the region taking off in this period (Sussman et al. 1994), farmers in Behompy started witnessing how after each episode of heavy rains, subsequent flooding and water recession, their *baiboho* fields were left covered with a layer of unfertile sand, whose thickness increased over the years.[10] Forest loss in these years can be concretely observed in the most accessible areas along the major communication axis in the region, the *Route Nationale* (RN, National Road) 9 connecting the regional capital with the population centres northwards along the coast (Blanc-Pamard 2009; Seddon et al. 2000), and the RN 7, linking Toliara with the central highlands to the Northeast (Samisoa 2001; Ranaivoson 2001). On both axes forest loss and degradation have been, respectively, linked to *hatsake* and to charcoal production, the latter to supply Toliara, a city experiencing one of the fastest growth rates in the country since the 1970s (Mana et al. 2001; Randriamanarivo 2001; Hoerner 1990; Raison 2000). For the specific case of Behompy, the rapid forest retreat throughout these years on the Belomotse and Vineta plateau, which progressed northwards until just few hundreds of metres to the southern bank of the Fiherena river already in the early 1990s, might have had a crucial role on the process of the silting of the *baiboho* fields. However, and despite the deforestation that has on several occasions been suggested as related to increased soil erosion and intensification of flooding in the region (Salomon 1982; Olson 1984), to date no systematic study has been conducted to investigate the causal link between these phenomena that could allow for a more solid conclusion on this point.

To add to this regional socio-environmental panorama, in the late 1980s the joint effect of two development strategies at the national and the international level had unforeseen and long-lasting implications for communities in Behompy and for the south-west region of Madagascar in general. In December 1989 Malagasy president Admiral D. Ratsiraka released the *Code des Investissements* (Invest Law[11]), a market-liberalization measure aimed at attracting foreign companies to invest in Madagascar, entailing export-tax breaks and other fiscal advantages (Escande 1995). In parallel, the same year the European Union announced the POSEI, a strategy intended to foster the economic and social convergence of Europe's most distant territories with metropolitan standards (CE 2000). In the case of France and within the POSEIDOM[12] plan, this strategy focused principally on the promotion of Ile de la Réunion, the most remote French *département*, located in the Indian Ocean but just 1700 km by ship from the city of Toliara. The main goal was to exploit the livestock potential of this *département* by granting subsidies to the import of animal feed, concretely for the pork and poultry sectors. The combined advantages provided by the Malagasy and European policies drove SOPAGRI,[13] the subsidiary in Madagascar of the main Reunionese agricultural co-operative, to open a branch in Toliara to centralize the collection of feed, chiefly maize, to be shipped to La Réunion, to which end a silo was built near the port as the POSEIDOM funds started being released in 1992. Notwithstanding the unfavourable edaphic and climatic conditions, countless producers across the South-west participated in the maize cultivation boom, in a region considered to have been supplying at that time over half of that crop's production through *hatsake*, of which a large share was destined for export to La Réunion (Minten et al. 2006; Escande 1995; Fauroux 2000).

Adding more complexity, and in parallel to the intricate situation of the national political crisis of this period and the subsequent economic downturn (Razafindrakoto et al. 2014), at the time the export market started taking off, southern Madagascar was suffering the effects of one of the most severe droughts in the last decades (OFDA 1993). In a phenomenon repeated on several occasions in the history of southern Madagascar, the drought led to an acute famine in the southernmost regions of the island, triggering out-flows of migration northwards (Fenn and Rebara 2003). Many Tandroy and Mahafaly migrants engaged into *hatsake* throughout the west and south-west of the island in these years, facilitated by the existence at that time of the right of clearing forest as a way to gain access to land (Casse et al. 2004; Réau 2002). Furthermore, in February 1991 cyclone Cynthia hit the region, leaving dozens of casualties and thousands of hectares of agricultural fields destroyed (Fauroux 2002; Donguy 1992). Cynthia caused intense flooding throughout the Fiherena river area, reinforcing the phenomenon of silting of the *baiboho* fields and pushing the population in Behompy into a critical situation. Households in the commune turned to the regional forest authorities to demand permission to clear forest land with which to make up for the agricultural land spoiled by the floods, permission that was granted to some of the most affected households.[14] In the years following, new agricultural fields began to be opened up in the forests north of the river, previously an area relatively undisturbed. Notwithstanding the permits being allowed for a defined surface of forest land to be converted to agricultural use, some of the families receiving these permissions expanded their plots much beyond what was stipulated in the documents.[15]

In 2002 the maize export to Ile de la Réunion was virtually over, and results for stakeholders located on both sides of the commodity chain were disparate. While between 1990 and 2003 the production of poultry and pork in La Réunion rose by 92% and 51%, respectively, the benefits accruing to populations more directly engaged in maize production in south-western Madagascar during the period were scarce and ephemeral (Minten et al. 2006). Moreover, in the region much of the benefits yielded throughout the lifespan of the maize boom were captured by the *Karana* import-export companies, controlling the bulk of the trade (Fauroux 2000; Escande 1995). Despite the difficulties in quantitatively linking the maize exported with the forest loss in the region during these years, some 250,000 ha for the period 1990–2000 (ONE et al. 2013), the environmental consequences of the maize boom are deemed to have been significant. According to a conservative estimate based on declared imports, no less than 160,000 tons of maize were exported from Madagascar to La Réunion between 1988 and 1998, which, assuming an optimistic yield of four tons per hectare of forest, could have contributed to at least almost a fifth of the regional deforestation in the period (Minten et al. 2006).

The end of this market roughly coincided with the outbreak of the political crisis that during months was to keep the Malagasy economy paralysed, due to the conflict between the two main contenders to the presidential seat, D. Ratsiraka and M. Ravalomanana (Razafindrakoto et al. 2014). Even though is difficult to relate these national-level issues with the evolution of local socio-economic dynamics in Behompy, the political impasse, which led to a deep regression of most social and economic indicators in the country, is remembered by inhabitants in the commune as a turning point, when *hatsake* started rapidly spreading all over the forested area north to the Fiherena. In the next years and in spite of the quantity of maize being exported to La Réunion sharply dropping after the boom vanished, the capillary collection and marketing networks for agricultural products continued operating in the region to supply maize to regional, national and international markets (Minten et al. 2006; Réau 2002). The existence of this network providing a channel to drain the output of a crop that, at least in the case of Behompy, would still be predominantly produced through *hatsake*, continued constituting a strong incentive for households to engage into maize cultivation and further expand their upland swidden cultivation fields.[16] To an important extent, throughout these years customary landowners in the commune resorted to hired labour and sharecropping agreements to exploit their *hatsake* plots.[17]

Further Cyclones and Implementation of Ranobe PK 32 New Protected Area

In March 2004 Gafilo, the most intense cyclone ever occurring in the south-western Indian Ocean basin swept most of the island. More critically, at the end of January 2005, the consecutive landfall of cyclone

Ernest and tropical storm Felapi near Toliara city within a week caused again huge devastation, with nearly forty casualties and thousands of hectares of agricultural fields destroyed (WMO 2005). As happened in the case of Angèle and Cynthia, the occurrence of these cyclones marked the onset of a protracted dry spell that in this case lasted until the beginning of 2007.

Throughout all these years the pace of expansion of *hatsake* on the spiny forests north of the Fiherena river increased unabated, and the release of a detailed report inventorying its unique biological wealth and its endangered status (e.g. Frontier-Madagascar 2005) eventually led to the creation of a nature conservation scheme with the aim of halting the forest retreat trend in the area. The protection scheme was part of the Durban Vision, an ambitious initiative announced in 2003 by President Ravalomanana with the goal of expanding the protected areas' system on the island from 1.6 million ha up to 6 million ha within five years. The goal was to match the International Union for the Conservation of Nature (IUCN) recommendations of extending the protection status up to 10% of the country's surface, a task to be largely delegated to Malagasy and international environmental organizations, given the overwhelming nature of the endeavour. The Durban Vision, materialized in the expanded *Système des Aires Protégés de Madagascar* (SAPM, Protected Areas System of Madagascar), had a strong focus on the dry forests of southern and western Madagascar. This was up to then the less protected biome in the country and the one experiencing the fastest forest retreat (Virah-Sawmy et al. 2014; Du Puy and Moat 1998; Waeber et al. 2015). Concretely in the spiny forest of the south-west, one of the floristic domains with the highest proportion of endemic species on the island (Phillipson 1996), several of its most ecologically outstanding areas were selected to be established as New Protected Areas (NPAs), an intermediate step before a status of full protection would be conceded.

The forests enclosed by the Fiherena and the Manombo rivers and the Mozambique Channel, the most important for faunal biodiversity in the whole south-west (Gardner et al. 2009a, b), was one such an area. The biological wealth of the area, a *Site d'Intérêt Biologique* (Site of Biological Interest) since 1991 but without protection status, was since long acknowledged, and recommendations for the granting of protection status dated at least from 1983 (Domergue 1983; Rejo-Fienena 1995; Seddon et al. 2000). The NPA was proposed as a multiple-use protected area (category VI of the IUCN guidelines), to allow for the sustainable use of natural resources while guaranteeing biodiversity conservation, with concretely three different land uses envisaged for the area (WWF 2013). By decreasing order of extension, first catalogued were *Zones d'Utilisation Durable* (sustainable use zones), some two-thirds of the total surface. Here certain forms of sustainable forest exploitation could still be carried out, but no further conversion of forest to agricultural land was allowed. Second was the *Zone d'Occupation Contrôlée* (controlled occupation zones), areas within the forested zone with permanent settlements where agriculture was already being practised. And third, the remaining 13.5% of the surface which was determined as *Noyaux Durs* (strict conservation zones), chosen by their ecological relevance and reserved for strict conservation and research purposes. The project also included a range of alternative livelihood options intended to be provided to the local communities to compensate for the new conditions of restricted access to forest resources, the main ones support for and improvement of permanent agriculture techniques, and promotion of agroforestry and ecotourism initiatives.[18]

However, when implementation of the NPA under the name of Ranobe PK32 eventually took place in 2008, covering nearly 160,000 ha and including most of Behompy's commune surface, it suffered major flaws. First, largely stemming from the rush to execute the plan within the time span stipulated and the hegemony of the international conservation organizations on the implementation process (Corson 2014; Amelot et al. 2012), a conflation between *consultation* of communities affected and *sensibilisation* led to a lack of real participation of local populations affected in the planning of boundaries and zones of the protection scheme.[19] Second, concerning the governance of the NPA and given the endeavour of running the large protection scheme, its management was devised to be a shared responsibility between local communities, regional authorities and a collaborating external organization, in the case of Ranobe PK32 the WWF. For this end, institutional arrangements were completed with representatives of all eight rural communes affected by the implementation of the protection scheme and this led to the establishment of *Communautés de Base* (CO-BAs), clustered under a new inter-communal political entity. However, in the case of the forests north to the Fiherena river, it was considered that no strong customary institutions existed before the implementation of the NPA given that the area was largely inhabited until just few years before (Virah-Sawmy et al. 2014). This hindrance meant an unbridgeable gap to overcome for the institutions created. They thus did not have the means to effectively curb the ongoing *hatsake* expansion trend, and particularly to supersede the existing tenure arrangements between families in Behompy, in cases involving several tens of hectares.[20] A third and key shortcoming was the lack of funding with which to address the daunting task of providing alternative livelihoods for the nearly 90,000 people living around the NPA and in multiple small hamlets scattered across the forested area, which made this goal unachievable in practice. Moreover, the withdrawing of international donors' aid to the country following the *coup d'état* in 2009 did not ease the situation either, meaning that eventually little of those alternatives materialized. Despite the implementation of the NPA and the expansion of its surface in 2010, in the following years *hatsake* continued spreading unabated throughout the forests of Behompy. Placing more strain on the agricultural activities of these rural populations, locust invasion, an old-known plague in the island and whose spread has been related to deforestation in other regions of the country (Lecoq et al. 2006; Franc et al. 2008), would become practically chronic to southern Madagascar from 2012 onward (FAO 2015).

Haruna

Adding tension to the situation, Cyclone Haruna, the strongest cyclone in living memory in this area, struck the south-western coast of Madagascar on 22 February 2013. It brought winds of up to 190 km/h and discharged over 200 mm of rainfall within 24 hours, making the Fiherena once again flood Toliara, as the dykes protecting the city from the waters collapsed under the pressure of the deluge. Twenty-six people lost their lives, more than 13,000 were left homeless and thousands of hectares of agricultural land destroyed just weeks before the harvest of several crops was expected. It triggered a humanitarian disaster in an already food insecurity-stricken area (WFP et al. 2013; ACF 2013). In Behompy, up to two-thirds of the 650 ha of *baiboho* available to inhabitants were severely affected and even in some villages nearly all permanent agriculture land surface in the *fokontany*.[21] A layer of sand, in cases up to one metre thick, was left on the fields after the floods receded, leaving the plots useless for cultivation. Eight months after the passage of Haruna, when this research was conducted, communities in Behompy were struggling to cultivate some tuber crops on the sand-covered plots. In other zones of the commune the *baiboho* disappeared altogether, as the river changed its course during the violent floods and came to run through the place where the plots were formerly located. As a result, a number of individuals were suddenly deprived of their means of living by the cyclone and were forced to leave their dwellings near the river banks, settling on the forested areas uphill to try to build a new life.[22]

Parallel to the great stress on existing agricultural practices caused by Haruna, the other means of livelihood available to inhabitants in Behompy came also under pressure for other reasons. As a consequence of the general context of political instability reigning in the country in the past years and particularly due the dramatic shrinkage of the state's budget after the coup of 2009 (due to cuts in foreign aid), the Malagasy state was severely hampered in its ability to enhance rural security in large areas of the country (Ploch and Cook 2012; McConnell et al. 2015). This resulted in an increase of rural insecurity, intensely felt in the south of the island. The most obvious indications of this situation are the spread of assaults on villages and cattle theft, perpetrated by *malaso* (rural bandits), which have increased both in frequency and death toll (Razafindrakoto et al. 2014; Pellerin 2014). This context of insecurity, already on the rise in the area of Behompy for decades (ILO 2001; Tahirindraza 2006; Hoerner 1990), has in recent years become an almost insurmountable burden for inhabitants. Within the panorama of shrinking livelihood options, and although it could constitute a potential alternative to resort to, to make up for the losses in sedentary agricultural activities, livestock rearing is also under increasing pressure.

Indeed, insecurity was the most significant problem highlighted by inhabitants in the commune at the moment fieldwork was carried out. Particularly compelling is the fact that most household heads surveyed in Behompy expressed their reluctance to engage into raising zebu, or to build up their herd again for fear of it being stolen, should the insecurity prevailing not decrease. Under such a climate of growing incidence of cattle thefts and as recently found for communes next to Behompy (Gardner et al. 2015), it could be not too adventurous to suggest that this phenomenon might be already driving households to abandon cattle rearing and, conceivably, to engage into more forest-based livelihoods, like *hatsake* or charcoal production.

Conclusion

The case of Behompy has served to bring to light aspects seldom taken into account when analysing factors shaping the context under which livelihood decisions are taken by populations in rural Madagascar. Furthermore, and despite the unique situation in this rural commune, many of the socio-economic and environmental aspects referred to in this chapter are not exclusive to Behompy. Rural insecurity is still on the rise in much of the island's southern regions, and the initiatives deployed by the Malagasy government to date, besides having important implications for human rights, are not showing much success in halting this trend. Furthermore, southern Madagascar is in the midst of a dry spell lasting for over half a decade now, the full social and environmental implications of which are still difficult to assess. To add to this, the most comprehensive assessment for future climate change scenarios conducted to date for Madagascar suggests that the southern regions of the island will experience further pressures (Tadross et al. 2008). Concretely, while cyclones might not increase in frequency, they are expected to gain in intensity. In addition, sharp rises in temperature are forecast for the middle of the present century, continuing the trend appreciable since the 1960s. For communities dwelling near the Fiherena river, and for rural populations of these regions of Madagascar in general, these scenarios might further constrain their ability to pursue sustainable livelihoods, with the subsequent implications for environmental change.

In addition to the entangled socio-economic and environmental pressures explored in this chapter, and as expressed by participants of this study, population growth should not be overlooked. In fact, when household heads in Behompy were asked whether life was easier or more difficult nowadays than ten years before, exponential population growth was, together with insecurity and the effects of cyclones on agricultural land, the most cited reason for life being more difficult. At first reading, this observation tends to lend neo-Malthusian arguments a hand. However, demographic growth was only mentioned in a qualitative way by local populations, which makes it difficult to assess its exact contribution within the broader span of factors influencing land use decisions taken by rural communities.

Many of the aspects brought into this study have been rarely analysed systematically in scientific and technical literature on which interventions aimed at encouraging rural communities' sustainable practices and preserving Madagascar's forests are commonly based. If effective conservation of the biodiverse spiny forest is to be achieved, all these dimensions should be considered in order to design and implement feasible initiatives to support local populations and address the numerous challenges they face.

Notes

1. Translation by the author.
2. The regional capital of Atsimo Andrefana, literally south-west in Malagasy language.
3. *Fokontany* is the smallest administrative unit in Madagascar, involving one or more villages or hamlets.
4. Source: interviewee-a. (To preserve anonymity of the interviewees, individuals whose responses are used in this chapter are coded with a letter.)
5. Irrigation of the plots is achieved by deviating water from the river by digging out channels on the clayish soils of the *baiboho*, which need to be rebuilt after each rainy season.

6. Indeed, in several of the occasions that the yearly average surpassed the 500 mm threshold, it was due to the effect of a cyclone.
7. Precipitation data for 1970–2013 were obtained from the Meteorological Station near the Toliara Airport, some 20 km southwest from Behompy.
8. Interviewee-b.
9. Interviewee-c.
10. Interviewee-d.
11. Loi no. 89-026 du décembre 1989.
12. *Programme d'Option Spécifique à l'Élargissement et à l'Insularité des Départements d'Outre-Mer* (Programme of Options Specific to the Remote and Insular Nature of the French Overseas Departments).
13. *Société de Production, de Stockage et de Manutention des Produits Agricoles* (Company of Production, Storage and Handling of Agricultural Products).
14. Interviewee-e and interviewee-f.
15. Interviewee-g and interviewee-h.
16. Interviewee-i and interviewee-d.
17. Interviewee-j, interviewee-k and interviewee-f.
18. Interviewee-l.
19. Interviewee-l and interviewee-m.
20. Interviewee-c and interviewee-g.
21. This estimation is made by the author based on several field visits to areas affected in the commune and subsequent visual interpretation of very high-resolution satellite imagery, available for periods before and after the passage of cyclone Haruna.
22. Interviewee-n and interviewee-o.

References

ACF. (2013). *Accès à l'eau potable, amélioration de l'assainissement et sécurité alimentaire pour les populations sinistrées par le cyclone Haruna - Rapport Narratif Final*: Action Contre la Faim.

AfDB. (2015). *Projet de réhabilitation des infrastructures agricoles dans la région sud-ouest (PRIASO)*. Retrieved May 4, 2015, from http://www.afdb.org/fr/projects-and-operations/project-portfolio/project/p-mg-aab-004/.

Amelot, X., Moreau, S., & Carrière, S. M. (2012). Des justiciers de la biodiversité aux injustices spatiales. L'exemple de l'extension du réseau d'aires protégées à Madagascar. In D. Blanchon, J. Gardin, & S. Moreau (Eds.), *Justice et Injustices Environnementales* (pp. 193–214). Nanterre: Presses Universitaires de Paris-Ouest.

Blanc-Pamard, C. (2009). The Mikea Forest under threat (southwest Madagascar): How public policy leads to conflicting territories. *Field Actions Science Report (Online), 3,* 1–12.

Brimont, L., Ezzine-de-Blas, D., Karsenty, A., & Toulon, A. (2015). Achieving conservation and equity amidst extreme poverty and climate risk: The Makira REDD+ Project in Madagascar. *Forests, 6*(3), 748.

Brinkmann, K., Noromiarilanto, F., Ratovonamana, R. Y., & Buerkert, A. (2014). Deforestation processes in south-western Madagascar over the past 40 years: What can we learn from settlement characteristics? *Agriculture, Ecosystems & Environment, 195,* 231–243.

Brown, M. L. (2009). Madagascar's cyclone vulnerability and the global vanilla economy. In E. C. Jones & A. D. Murphy (Eds.), *The political economy of hazards and disasters* (pp. 241–264). Plymouth: AltaMira Press.

Burivalova, Z., Bauert, M. R., Hassold, S., Fatroandrianjafinonjasolomiovazo, N. T., & Koh L. P. (2015). Relevance of global forest change data set to local conservation: Case study of forest degradation in Masoala National Park, Madagascar. *Biotropica 47*(2), 267–274.

Casse, T., Milhøj, A., Ranaivoson, S., & Randriamanarivo, J. R. (2004). Causes of deforestation in southwestern Madagascar: What do we know? *Forest Policy and Economics, 6*(1), 33–48.

CE. (2000). Evaluation de l'impact des actions realisées en exécution du volet agricole du POSEIDOM. Retrieved November 9, 2016, from http://ec.europa.eu/agriculture/eval/reports/poseidom/index_fr.htm.

Chaperon, P., Danloux, J., & Ferry, L. (1993). *Fleuves et Rivières de Madagascar. Ony Sy Renirano Eto Madagasikara.* Paris: ORSTOM.

Corson, C. (2014). Conservation politics in Madagascar: The expansion of protected areas. In I. R. Scales (Ed.), *Conservation and environmental management in Madagascar* (pp. 193–215). Abingdon and New York: Routledge.

de Saint-Sauveur, A. (1996). Le vol de bétail, facteur clef de l'organisation foncière et pastorale dans le pays bara (Sud-Ouest malgache). *Journal d'Agriculture Traditionnelle et de Botanique Appliquée, 38*(2), 253–267.

Domergue, C. (1983). Note préliminaire en vue de la mise en réserve de la forêt du point kilomètre 32 au nord de Tuléar Madagascar. *Bulletin de l'Académie Malgache, 61,* 105–114.

Donguy, P. (1992). Phénomènes Cycloniques dans l'Océan Indien vus par Météosat (Saison 1990–1991). Veille Climatique Satellitaire (pp. 79–97). Paris: Ministère de la Coopération, Meteo France, ORSTOM.

Du Puy, D. J., & Moat, J. F. (1998). Vegetation mapping and classification in Madagascar (using GIS): Implications and recommendations for the conservation of biodiversity. In C. R. Huxley, J. M. Lock, & D. F. Cutler (Eds.), *Chorology, taxonomyn and ecology of the Floras of Africa and Madagascar* (pp. 97–117). Kew: Royal Botanic Gardens.

Escande, C. (1995). Étude des Réseaux Commerciaux et de la Formation des Prix des Produits Agricoles dans le Sud-Ouest de Madagascar. Montpellier: CNEARC Montpellier. ENSAIA Nancy. DEA Thesis, 117.

FAO. (2013). Crop water information: Maize. Retrieved July 2, 2015, from http://www.fao.org/nr/water/cropinfo_maize.html.

FAO. (2015). *Bulletin de Situation Acridienne Madagascar. Bulletin de la première décade de juin (2015-D16)*. Rome: Food and Agricultural Organization of the United Nations.

Fauroux, E. (1989). Boeufs et pouvoirs: les éleveurs du Sud-Ouest et de l'Ouest malgaches. *Etats et Sociétés Nomades, 34,* 63–73.

Fauroux, E. (2002). Les sociétés rurales de l'Ouest malgache: des transformations profondes et complexes. *Afrique Contemporaine, 202–203,* 111–132.

Fauroux, S. (2000). Instabilité des cours du maïs et incertitude en milieu rural: le cas de la déforestation dans la région de Tuléar (Madagascar). *Tiers-Monde, 41*(164), 815–839.

Fenn, M., & Rebara, F. (2003). Present migration tendencies and their impacts in Madagascar's spiny forest ecoregion. *Nomadic Peoples, 7*(1), 123–137.

Franc, A., Soti, V., Tran, A., Leclair, D., Duvallet, G., & Duranton, J.-F. (2008). Deforestation, new migration pathways and outbreaks of the Red locust Nomadacris septemfasciata (Orthoptera: Acrididae) in the Sofia river basin. Paper Conférence SDH-SAGEO. Montpellier, 23–26 juin 2008, 1–17.

Frontier-Madagascar (Thomas, H., Kidney, D., Rubio, P., & Fanning, E.) (Eds.). (2005). *The Southern Mikea. A Biodiversity Survey. Toliara: Frontier-Madagascar Environmental Research Report 12.* Society for Environmental Exploration, UK, and Institut Halieutique et des Sciences Marines.

Ganzhorn, J. U. (1995). Cyclones over Madagascar: Fate or fortune? *Ambio, 24*(2), 124–125.

Gardner, C. J., Fanning, E., Thomas, H., & Kidney, D. (2009a). The lemur diversity of the Fiherenana—Manombo Complex, southwest Madagascar. *Madagascar Conservation and Development, 4*(1), 38–43.

Gardner, C. J., Kidney, D., & Thomas, H. (2009b). First comprehensive avifaunal survey of PK32-Ranobe, a new protected area in south-western Madagascar. *Phelsuma, 17,* 20–39.

Gardner, C. J., Gabriel, F. U. L., St. John, F. A. V., & Davies, Z. G. (2015). Changing livelihoods and protected area management: A case study of charcoal production in south-west Madagascar. *Oryx 50*(3), 495–505.

Grouzis, M., & Milleville, P. (2001). 'Modèle d'analyse de la dynamique des systèmes agro-écologiques'. Sociétés paysannes, transitions agraires et dynamiques écologiques dans le sud-ouest de Madagascar. In S. Razanaka, M. Grouzis, P. Milleville, B. Moizo, & C. Aubry (Eds.), *Actes de l'Atelier CNRE-TRD, novembre 1999* (pp. 229–238). Antananarivo: CNRE-TRD.

Harvey, C. A., Rakotobe, Z. L., Rao, N. S., Dave, R., Razafimahatratra, H., Rabarijohn, R. H., et al. (2014). Extreme vulnerability of smallholder farmers to agricultural risks and climate change in Madagascar. *Philosophical Transactions of the Royal Society B: Biological Sciences, 369*(1639), 20130089.

Hochrainer-Stigler, S., Mechler, R., & Mochizuki, J. (2015). A risk management tool for tackling country-wide contingent disasters: A case study on Madagascar. *Environmental Modelling and Software, 72,* 44–55.

Hoerner, J. M. (1979). *Géographie Régionale du Sud Ouest de Madagascar.* Collection TSIOKANTIMO, Série recherche no 5: Centre Universitaire Régionale de Tulear.

Hoerner, J. M. (1982). Les vols de boeufs dans le sud Malgache. *Madagascar Revue de Géographie, 41,* 85–105.

Hoerner, J. M. (1990). *La Dynamique Régionale du Sous-Développement du Sud-Ouest de Madagascar.* Montpellier and Perpignan: Université Paul Valéry.

ILO. (2001). Madagascar commune census. Cornell University, FOFIFA, INSTAT, USAID.

Kull, C. A. (2014). The roots, persistence, and character of Madagascar's conservation boom. In I. R. Scales (Ed.), *Conservation and environmental management in Madagascar* (pp. 146–171). Abingdon and New York: Routledge.

Lecoq, M., Franc, A., Luong-Skovmand, M.-H., Raveloson, A., & Ravelombony, V. D. P. (2006). Ecology and migration patterns of solitary red locusts, Nomadacris septemfasciata (Serville) (Orthoptera: Acrididae) in southwestern Madagascar. *Annales de la Société Entomologique de France (N.S.)/International Journal of Entomology, 42*(2), 197–205.

Lewis, R. J., & Bannar-Martin, K. H. (2012). The impact of cyclone Fanele on a tropical dry forest in Madagascar. *Biotropica, 44*(2), 135–140.

Mana, P., Rajaonarivelo, S., & Milleville P. (2001). 'Production de charbon de bois dans deux situations forestières de la région de Tuléar'. Sociétés paysannes, transitions agraires et dynamiques écologiques dans le sud-ouest de Madagascar. In S. Razanaka, M. Grouzis, P. Milleville, B. Moizo, & C. Aubry (Eds.), *Actes de l'Atelier CNRE-TRD, novembre 1999* (pp. 199–210). Antananarivo: CNRE-TRD.

Marcus, R. (2007). Where community-based water resource management has gone too far: Poverty and disempowerment in Southern Madagascar. *Conservation and Society, 5*(2), 202–231.

McConnell, W., Viña, A., Kull, C., & Batko, C. (2015). Forest transition in Madagascar's highlands: Initial evidence and implications'. *Land 4*(4), 1155. http://www.mdpi.com/2073-445X/4/4/1155/pdf. Accessed May 5, 2017.

Mercier, J.-R. (2006). The preparation of the National Environmental Action Plan (NEAP): Was it a false start? *Madagascar Conservation and Development, 1*(1), 50–54.

Mertz, O., Padoch, C., Fox, J., Cramb, R. A., Leisz, S. J., Lam, N. T., et al. (2009). Swidden change in Southeast Asia: Understanding causes and consequences. *Human Ecology, 37*(3), 259–264.

Milleville, P., Grouzis, M., Razanaka, S., & Bertrand, M. (2001). La culture pionnière de maïs sur abattis-brûlis (hatsaky) dans le sud-ouest de Madagascar. 2. Évolution et variabilité des rendements. In S. Razanaka, M. Grouzis, P. Milleville, B. Moizo, & C. Aubry (Eds.), *Actes de L'Atelier CNRE-TRD novembre 1999* (pp. 255–268). Antananarivo: CNRE-TRD.

Minten, B., Meral, P., Randrianarison, L., & Swinnen, J. F. M. (2006). *Trade liberalization, rural poverty and the environment: The case of Madagascar.* Antananarivo: WWF Madagascar.

Montagne, P., Razafimahatratra, S., Rasamindisa, A., & Crehay, R. (Eds.). (2010). *Arina, le charbon de bois à Madagascar: entre demande urbaine et gestion durable.* Antananarivo: CITE.

OFDA. (1993). *Significant data on major disasters worldwide. 1900-Present.* Washington, DC: Office of the United States Foreign Disaster Assistance, Agency for International Development.

Olson, S. H. (1984). The robe of the ancestors: Forests in the history of Madagascar. *Journal of Forest History, 28*(4), 174–186.

ONE, DGF, FTM, MNP, & CI. (2013). *Evolution de la Couverture de Forêts Naturelles à Madagascar 2005–2010.* Antananarivo.

Pellerin, M. (2014). *Madagascar. Gérer l'héritage de la transition*. Paris: IFRI.
Phillipson, P. B. (1996). Endemism and non-endemism in the flora of southwest Madagascar. In W. Lourenço (Ed.), *Biogéographie de Madagascar. Biogeography of Madagascar* (pp. 125–136). Paris: Éditions de l'ORSTOM.
Ploch, L., & Cook, N. (2012). Madagascar's political crisis. Paper, Congressional Research Service.
Raharimalala, O., Buttler, A., Ramohavelo, C. D., Razanaka, S., Sorg, J.-P., & Gobat, J.-M. (2010). Soil–vegetation patterns in secondary slash and burn successions in Central Menabe, Madagascar. *Agriculture, Ecosystems & Environment, 139*, 150–158.
Raison, J. P. (2000). Madagascar: vers une nouvelle géographie régionale. *L'Information Géographique, 64*(1), 1–19.
Ranaivoson, J. R. S. (2001). Fonciere et déforestation sur le plateau de Vineta. Sociétés paysannes, transitions agraires et dynamiques écologiques dans le sud-ouest de Madagascar. In S. Razanaka, M. Grouzis, P. Milleville, B. Moizo, & C. Aubry (Eds.), *Actes de l'Atelier CNRE-TRD novembre 1999* (pp. 39–51). Antananarivo: CNRE-TRD.
Randriamanarivo, R. (2001). L'activité charbonnière dans les économies paysannes (Axe Routierandranovory-Tuléar Rn7). Sociétés paysannes, transitions agraires et dynamiques écologiques dans le sud-ouest de Madagascar. In S. Razanaka, M. Grouzis, P. Milleville, B. Moizo, & C. Aubry (Eds.), *Actes de l'Atelier CNRE-TRD novembre 1999* (pp. 211–221). Antananarivo: CNRE-TRD.
Rasambainarivo, J. H., & Ranaivoarivelo, N. (2006). *Country pasture/forage resource profiles*. Madagascar and Rome: FAO.
Rasamimanana, H., Ratovonirina, A. J., & Pride, E. (2000). Storm damage at Berenty Reserve. *Lemur News, 5*, 7–8.
Ratsimbazafy, J. H., Ramarosandratana, H. V., & Zaonarivelo, R. J. (2002). How do black-and-white ruffed lemurs still survive in a highly disturbed habitat? *Lemur News, 7*, 7–10.
Razafimanantsoa, D. (1987). La commercialisation des bovides dans le Sud-Manombo. In M. Fieloux & J. Lombard (Eds.), *Étude des Transformations Socio-Economiques dans Le Sud-Ouest Malgache: L'Exemple du Couloir d'Antseva* (pp. 163–177). Antananarivo: MRSTD and Paris: ORSTOM.
Razafindrakoto, M., Roubaud, F., & Wachsberger, J.-M. (2014). *Jalons pour une économie politique de la trajectoire malgache: une perspective de long terme*. Paris: Université Paris-Dauphine, UMR DIAL and IRD.

Réau, B. (2002). Burning for zebu: The complexity of deforestation issues in western Madagascar. *Norsk Geografisk Tidsskrift – Norwegian Journal of Geography, 56,* 219–229.

Rejo-Fienena, F. (1995). *Etude Phytosociologique de la Végétation de la Région de Tuléar (Madagascar) et Gestion des Ressources Végétales par les Populations Locales (Cas du P.K. 32)* (PhD thesis, Muséum National d'Historie Naturelle, Toliara).

Ruthenberg, H. (1971). *Farming Systems in the Tropics.* Oxford: Clarendon Press.

Salomon, J.-N. (1982). Realités et conséquences de la déforestation dans l'Ouest Malgache. *Madagascar Revue de Géographie, 40,* 7–13.

Samisoa. (2001). Migrations et Déforestation sur le Plateau Calcaire de Belomotse-Vineta. Sociétés paysannes, transitions agraires et dynamiques écologiques dans le sud-ouest de Madagascar. In S. Razanaka, M. Grouzis, P. Milleville, B. Moizo, & C. Aubry (Eds.), *Actes de l'Atelier CNRE-TRD novembre 1999* (pp. 53–62). Antananarivo: CNRE-TRD.

Scales, I. R. (2011). Farming at the forest frontier: Land use and landscape change in Western Madagascar, 1896–2005. *Environment and History, 17,* 499–524.

Scales, I. R. (2014). The drivers of deforestation and the complexity of land use in Madagascar. In I. R. Scales (Ed.), *Conservation and environmental management in Madagascar* (pp. 103–125). Abingdon and New York: Routledge.

Seddon, N., Butchart, S., Tobias, J., Yount, J. W., Rémi Ramanampamonjy, J., & Randrianizahana, H. (2000). Conservation issues and priorities in the Mikea Forest of south-west Madagascar. *Oryx, 34*(4), 287–304.

Sussman, R. W., Green, G. M., & Sussman, L. K. (1994). Satellite imagery, human ecology, anthropology, and deforestation in Madagascar. *Human Ecology, 22*(3), 333–354.

Tadross, M., Randriamarolaza, L., Rabefitia, Z., & Yip, Z. K. (2008). *Climate change in Madagascar; recent past and future: Climate Systems Analysis Group, University of Cape Town.* Antananarivo: South Africa and National Meteorological Office Madagascar.

Tahirindraza, H. S. (2006). *Monographie du village Behompy-Mahasoa* (Master's thesis, Université de Toliara).

Urech, Z. L., Zaehringer, J. G., Rickenbach, O., Sorg, J.-P., & Felber, H. R. (2015). Understanding deforestation and forest fragmentation from a livelihood perspective. *Madagascar Conservation and Development, 10*(2), 67–76.

Virah-Sawmy, M., Gardner, C. J., & Ratsifandrihamanana, A. N. (2014). The Durban Vision in practice. Experiences in the participatory governance of Madagascar's new protected areas. In I. R. Scales (Ed.), *Conservation and environmental management in Madagascar* (pp. 216–251). Abingdon and New York: Routledge.

van Vliet, N., et al. (2012). Trends, drivers and impacts of changes in swidden cultivation in tropical forest-agriculture frontiers: A global assessment. *Global Environmental Change, 22*, 418–429.

von Heland, J., & Folke, C. (2014). A social contract with the ancestors—culture and ecosystem services in southern Madagascar. *Global Environmental Change, 24*, 251–264.

Waeber, P. O., et al. (2015). Dry forests in Madagascar: Neglected and under pressure. *International Forestry Review, 17*(S2), 127–148.

Waeber, P. O., Wilmé, L., Mercier, J.-R., Camara, C., & Lowry, P. P., II. (2016). How effective have thirty years of internationally driven conservation and development efforts been in Madagascar? *PLoS ONE, 11*(8), e0161115.

WFP, MinAgri, ACF, FAO, CARE, FID, ONN, BNGRC, ICCO, GRET, CRS, RTM, ADRA, Caritas, SAF, WHH, & Land-O'Lakes. (2013). *Evaluation des Impacts du Cyclone Haruna sur les Moyens de Subsistance, et sur la Sécurité Alimentaire et la Vulnérabilité des Populations Affectées*. Antananarivo: n.p. http://documents.wfp.org/stellent/groups/public/documents/ena/wfp257764.pdf?iframe. Accessed April 04, 2017.

WMO. (2005). *RA I tropical cyclone committee for the south-west Indian ocean seventeenth session*. Gaborone: World Meteorological Organization.

World Bank. (1988). *Madagascar—Environmental action plan*. World Bank – USAID - Cooperation Suisse – UNESCO – UNDP - WWF.

World Bank. (1995). *Madagascar—Irrigation rehabilitation project*. Washington, DC: World Bank.

World Bank. (1996). *Madagascar poverty assessment*. Volume II: Main Report, Annexes. Washington, DC: The World Bank.

World Bank. (2003). *Madagascar. Rural and environment sector review*. Volume I: Main Report. Washington, DC: The World Bank.

World Bank. (2015). *Madagascar—Systematic country diagnostic*. World Bank Group.

WWF. (2013). *Nouvelle Aire Protégée Ranobe PK32. Plan d'Aménagement et de Gestion 3 – Plan d'Aménagement et de Zonage*. Mangily, Madagascar: Mamisoa Andriafanomezana and World Wildlife Fund.

10

Challenging Impediments to Climate Change Initiatives in Greg Mbajiorgu's *Wake Up Everyone*

Norbert Oyibo Eze

Introduction

This chapter is a study of the literary imagination of environmental problems in Africa, particularly in the Niger Delta region of the Federal Republic of Nigeria. With its very apt title, *Wake Up Everyone*, by Nigerian playwright, monodramatist and academic, Greg Mbajiorgu, is an artistically bold and challenging work which, according to Nicolas Ozor, "provides one of the effective strategies (drama) for communicating and socializing the science of climate change especially at the grass root level" (From the blurb). This play that Chimalum Nwankwo adjudges successful both as an "exciting entertainment and intellectual adventure" (from the blurb) underscores the need for a multidisciplinary approach in handling the subject of climate change. The play not only showcases the causes and effects of climate change in Nigeria, but also discusses its politics and even advocates workable strategies for

N. O. Eze (✉)
University of Nigeria, Nsukka, Nigeria

© The Author(s) 2018
J. Abbink (ed.), *The Environmental Crunch in Africa*,
https://doi.org/10.1007/978-3-319-77131-1_10

challenging the impediments to climate change initiatives in the country. In fact, the theme song of the play, "wake up, wake up everyone", is a resounding advocacy call which, according to Damian Opata (2014), suggests that "… sustaining a balanced and productive environment is an insistent social obligation for all of humanity" (from the blurb).

In this study, however, an attempt will be made through textual analysis and interpretation not to consider its artistic-literary merits but to explain what the play considers the primary issues that must be addressed if the efforts towards climate change adaptation and mitigation must yield fruits in Africa. Before going on to do this, it is pertinent to give a summary of the play.

About the Play

Wake Up Everyone (first published in 2011) is written in three Acts. The play is set in Ndoli land, a coastal town in the Niger Delta region of Nigeria. In the play, Chief Edwin Ochonkeya is presented as exploiting the massive oil spillage in the area to get huge sums of money from the oil companies whose activities caused the spillage. He does this through litigation. He equally used the chance to get the companies to bankroll his election into the position of the Local Government Chairman. Nevertheless, his poor political orientation and squandermania culture did not allow him to use his position to assist Professor Aladinma, an environmental vanguard activist, in his efforts to devise proactive measures to check an impending flood. His refusal to bring counterpart funding to support the initiative of the UNDP makes the building of dykes proposed by the Professor impossible. Eventually the flood comes, leading to overwhelming destruction of farmlands, and this provoked a community-wide violent demonstration against the Local Government Chairman. Indeed, oil exploration and exploitation in the Niger Delta area of the South South geopolitical zone of Nigeria as well as the politics associated with it have continued to unleash disheartening disasters on the people and their environment. The play X-rays the complex realities and unhealthy social relations which make the Niger Delta an abode of conflicts and all kinds of casualties. Obi (2008, 12) observes that, "The swamps and creeks of the Niger Delta hold deep in their bowels the irresistible allure of petroleum and natural gas: the black and blue gold-the very stuff that drives the engines and wheels of modern living". The reckless exploitation of these resources and the utter degradation of the environment and neglect of the people, by the multinational capitalists and their local cronies, turn this place into what Ossie Enekwe (1990) describes as. "Rich Land, poor people". Reinforcing this paradoxical perception of the Niger Delta, Obi (2008, 12–13) writes:

> A living paradox of wealth amidst poverty, rural locals hosting the world's richest and globally integrated multinationals, fuelling the world but hardly having enough for its own villages, polluted creeks and poisoned rivers, teeming with youth but no jobs, peopled by ethnic minorities in a federal system where power and resources are centralized. And dominated by ethnic majorities, hosting the life blood of the Nigerian state, watching it being piped away. This delta is immersed in… conflicts in which hundred of workers from over a dozen countries have been kidnapped and mostly ransomed and released, casualties recorded on the part of the 'militants' and security forces as well as villagers caught in the cross-fire.

Wake Up Everyone is seen by Eze (2013, 67) as a dramatic tour de force on the subject of climate change, which itself is arising from the capitalists' mindless exploitation of oil in the Niger Delta and the intricate social realities emanating from it.

Analysis and Interpretation

In Act one, scene one of the play, Professor Aladinma, a major character in the play and a strategist for climate change adaptation and mitigation, works hard to convince the Local Government Chairman that climate change is not a realization of the predictions of "the biblical last days", but a planetary reaction "to man's mindless activities". Here, the playwright aspires to puncture ignorance and mythopoeic assumptions, which are frequently encountered in Africa with respect to certain issues like climate change. Mythical thinking is still prevalent in rural Africa, and it is given weight in the play in the discussion between the local farmers–Anayo and Mazi Chinedum where they lament poor agricultural yields and the problems of fishing:

ANAYO	Terrible, in fact, I lack words to explain how disappointing the earth is this season. The size of yam and cassava are ridiculous. Look at those ones (Takes some tubers from his bicycle). What do you call these? Sticks or tubers? I was wondering if I could follow you to the river to learn how to fish. Obviously fishing will be a bit laborious with the high water level but not as strenuous and unrewarding as farming nowadays.
MAZI CHINEDUM	What are you talking about? What do you think is the cause of my wife's hullabaloo? For several days now, I've been to the river without any success or have you not noticed how expensive fish has become in the market recently…
ANAYO	Yes, something is wrong, something is really wrong. The earth is barren, our rivers that used to be filled with fishes and periwinkles are now empty, yet the rain will not stop and the cold…by the gods, the cold is terrible. All my children are sick with cold.
MAZI CHINEDUM	But these are the things Prof. Aladinma was trying to explain in our last meeting with him…
ANAYO	Forget Aladinma and all his shit about climate change. There is nothing like climate change, it is just that the gods are angry with us. (Act Three, scene two)

From Anayo's perspective, "Someone must have done something that provoked the gods". For him, the solution to this untoward happening lies in "placating the gods". Mythopoeic explanation of climate change is very endemic in Africa. Ezeliora (2000, 46) notes that, "Across the regions of Africa it is observed that man's quotidian experiences are spiritually tied to a Supreme Being through the links of the ancestors and other beings who are believed to protect the living from the world beyond". Although Akwanya (2004, 127) argues that myth is not an exclusive primitive mode of behaviour since "the capacity of myth-making is part of the make up of every mind", *Wake Up Everyone* supports the idea that, in the case of Niger Delta, the "religious imperatives have been overwhelmed by the current realities of economic and political problems" (Ezeliora 2000, 46). The text locates human experiences in the region within history that is directly pertinent to the people, and as Ezeliora (2000, 47) would state, "within the ambient of modern technology and the limitations imposed on humanity by the modern trends of mercantilism, avarice, arrogance and political gerrymandering". In fact, in Lenin's opinion, as cited in Opata and Ohaegbu (2000, 85), the play attempts to "cognize and problematize the contradictions and alienation in human and social relationships emanating from bourgeois ethics and psychology". *Wake Up Everyone* disentangles the problem of climate change from any form of supernatural influence, by placing the whole burden of the subject squarely on human shoulders. According to the text:

> Our soil and rivers have become unproductive because of chemicals and oil spillages. The floods and erosions we experience are caused by our senseless attempts to reclaim wetlands. What about the carbon monoxides from power generators and poorly maintained automobiles, and the unfriendly substances flared up into the air by oil companies on daily basis? (Act One, scene one).

The above statement is a critique of capitalism and industrialization. It suggests that the things modern people do for living are "destructively symptomatic" (Kahn 2010, 35). Foster's views (cited in Richard Kahn's *Critical Pedagogy, Ecoliteracy and Planetary Crisis* 2010, 3) give support to the playwright's point of view: "Over the last fifty to sixty years, then, a particularly noxious economic paradigm has unfolded like a shock wave across the face of the earth, one that has led to an exponential increase of global capital and startling achievements in science and technology, but which has also had devastating effects upon ecosystems both individually and taken as a whole." Before the discovery of crude oil at Oloibiri in a commercial quality in 1956, the Niger Delta was only mentioned in global history in relation to the trans-Atlantic slave trade, when it served as an outlet for slaves heading to America, Europe and the West Indies. However, the discovery of oil led to the scrambling for this area by the capitalist multinational oil drilling and marketing companies. According to Obi (2008, 12), "Running under villages, waters, mangrove forests, farms and sacred places are pipes like veins from which oil and gas are pumped daily-feeding the growing, almost insatiable appetite of the developed and more prosperous world". The activities of the greedy oil firms have continued to degrade the people's land, air, water and aquatic lives. Okpoko (2004, 106) states that "The damage has brought untoward economic, as well as socio-cultural and psychological hardships upon the people".

The implication is that climate change is an outgrowth of reckless human economic activities. The playwright seems to be of the opinion that since climate change results from collective aggression against nature, its mitigation requires collective human efforts. However, in the play, unlike the peasant farmers who are beginning to yield to the sensitization work of Professor Aladinma by subscribing to improved seeds and new farming techniques, the Local Government Chairman combines superstitious belief and poor political culture to undermine Professor Aladinma's effort of devising proactive measure against the looming flood. The Chairman sees Professor Aladinma's proposal as a kind of ivory-tower theory, something that may not be field-tested. He tells the Professor, that this fear of impending flood is what he cannot understand: "I'm aware that this is a riverine area and that we are surrounded by water. But since I was born, this community has never experienced flood; neither have my parents told me of any in their own days" (Act One, scene one). The above presents a "conservative" position, which seems to be prevalent in many developing countries (but not exclusively). This idea of not making provisions for change is countered by the Professor, who argues that empiricism—appraisal of the facts on the ground—should be the deciding factor, not tradition.

He reminds the Chairman that culture is dynamic. For him, "we must learn from other people's experiences" (Act One, scene one).

The Professor argues also that the notion of things remaining as they are no longer holds in a world driven by science and technology. His argument that Ndoli, as a coastal town, is susceptible to flooding is tenable because other coastal areas in the country have experienced it already. This is not only because of the town's closeness to the sea but also because of the undue pressure which oil drilling put on the sea. In general, as Baz Kershaw argues, "The threat of unchecked pollution facing Earth's biosphere, a possible calamity for humanity, is adding a further twist to its future" (2007).

Again, apart from having to run battle with issues of conservatism, superstition and ignorance, the play suggests that a major fight must be waged against poor political culture and orientation as well as a self-aggrandizing system where leaders and administrators go for personal gains and power. Chinua Achebe (1983, 1) has argued that, "The trouble with Nigeria is simply and squarely a failure of leadership". According to him, the Nigerian problem is "the unwillingness or inability of its leaders to rise to the responsibility, to the challenges of personal example, which are the hallmarks of true leadership" (1983, 1). He traces the foundation of poor leadership in the country to the founding fathers of the nation-Nnamdi Azikiwe who is noted to have said "that henceforth I shall utilize my earned income to secure my enjoyment of a high standard of living" and to Obafemi Awolowo who is equally reported to have said, "I was going to make myself formidable intellectually, morally invulnerable, to make all the money that is possible for a man with my brains and brawn to make in Nigeria". For Achebe (1983, 11–12), thoughts such as these are more likely to produce "aggressive millionaires" than selfless leaders of their people. As he writes: "An absence of objectivity and intellectual rigour at the critical moment of a nation's formation is more than an academic matter. It inclines the fledging state to disorderly growth and mental deficiency." Here, Achebe is right, for Nigeria has been very unfortunate with its leadership. In fact, since independence, it has been a train of leaders that promote ethnic bigotry and parochialism, hypocrisy, social injustice and manipulation, as well as, what Okolo (1994) terms "squandermania mentality". The squandermania mentality concept "is that disposition in a people by which they conceive and judge things mostly in terms of their consumable value" (9). Okolo (1994, 9) further argues that this habit of living "narrows down people's horizon by defining and determining progress, social importance, social values, power, authority, etc. almost exclusively in terms of material success and achievement". Squandermania culture is marked by "unproductive spending or wealth exhibitionism" (9). Enekwe (1990, 155) laments bitterly about this failure of the Nigerian political elite:

> The Nigerian elite class is not creative. It is not patriotic. It has its soul in London and New York…It steals the wealth of the nation and transforms it into nothing. By serving foreign interests, our elite class has ensured that Nigerians will remain unable to manage their affairs in the future, thus perpetuating poverty despite the incredible richness of our land.

Young (1997, 3) argues that "Governance arises as a matter of public concern". According to him, "[i]t emerges as a basis for cooperation; however, when opportunities arise to enhance social welfare by acting to co-ordinate the activities of individual members of the group" (p. 4). The action of the Chairman in *Wake Up Everyone* negates and subordinates social welfare to unprofitable philandering and profligacy. The Chairman opposes the idea of climate change and believes that spending money on the mitigating projects is unwarranted. Remaining adamant to the calamity that threatens his people is a major concern of the play and one that must be redressed if the dream of saving the people from planetary crisis can be actualized.

The Chairman of the local government represents the type of characters that parade much of the African political landscape. He is a symbol of an unsettling political orientation and culture of "governance" that has continued to prevent genuine, accountable development in Africa. Unlike Professor Aladinma, who is an "empowering representation of the marginalized" (*Theatre Research International*, 1), the Chairman demonstrates sufficient evidence of being a self-serving politician. He argues with the Professor by telling him that he appreciates his sensitization work on the issue of climate change adaptation, but when the Professor requests him to supply counterpart funding needed for the climate change mitigation work and to get the oil companies to assist him in that direction, he quickly retorts that the local government headquarters was recently refurbished by Zodiac Oil, the three eighteen-seater buses and two Toyota Hilux trucks out there were donated by the Continental Petroleum, the official Prado Jeep and the two hundred KVA soundproof generating set that is powering this local government Secretariat came three weeks ago as birthday gifts from the MD of Diamond Oil, ".. and", he says, "don't forget all these oil companies came together to raise me a loan with which I ran the election for this position. I have not begun to think of how to pay back the money and you are asking me to go back to them plate in hand like Oliver Twist, asking for more?" (Act One, scene one).

The above reveals the widespread image of politicians as being driven by the spirit of primitive acquisition or private rent seeking. It is glaringly obvious that the self-serving tendency of politicians pushes them to pursue wealth with reckless abandon, to the neglect of the well-being of the people who elected them into office. Eze (2013, 70) posits that "Every element which money is spent on in Chairman's speech, is a token of status symbol and ego massage, a means of showcasing and entrenching the self. Not a single item there serves the interest of the people". In a 2015 lecture entitled, *Rebuilding the Nigerian Dream* (2015), Mr. Osita Chidoka, the Nigerian Minister of Aviation, stated that:

> Total federal allocation to Local governments across the country in 2012 totalled N1.5 trillion-a figure that has been increasing year on year… However, most local governments dedicate the bulk of their allocation to recurrent expenditure of salaries and administrative cost rather than on capital projects such as education, health, public maintenance.

In the play, the Chairman of the local government illuminates this idea when he tells the Professor that "seventy percent of local government allocation is spent on workers' salaries while the remaining thirty percent is all that is left to run the local government" (Act One, scene one).

What is pathetic is that while the Chairman sees the Professor's proposal as something costly and unacceptable, he gladly tells his friend Jango, "If you are ready to help me milk this land, join me for a drink"(Act One, scene one). This portrays the Chairman as a self-aggrandizing politician, a fact which he further demonstrates vividly in Act Three, scene one, where he prefers the company of a seductive lady to a discussion of the Professor's proposal. The issue of our leaders fiddling with the people's collective destiny constrains the Professor to tell Desmond that "[i]n this country, we have done nothing, absolutely nothing; our leaders are still where Karl Marx's definition left them; we are still in our sorry state of primitive acquisition. Our leaders who should show more concern are completely nonchalant, crisis is not their business. The human condition is nowhere on their horizon." (Act Two, scene three).

From Professor Aladinma's lamentation above, it is clear that the idea of what to do is there, but money, the deciding factor, is at the command of politicians who are not willing at all to channel it to projects that will benefit the people. The playwright's argument, therefore, is that, "Government must put together a planned process and take deliberate steps to create institutions and structures that will stimulate the adaptive capacity of rural dwellers to the threat of climate change"(Act Three, scene one). He seems to agree with Ninimmo Bassey (2013, 3) that our "politicians should halt fiddling while the planet burns". He argues that so far, "this country is asleep" (Act Three, scene one) and that while leaders in developed world are proactive, "building quake resistant houses…, houses that can float on water", etc., leaders in Nigeria are only busy stuffing the cities with hotels, filling stations, shopping malls, exotic mansions and no breathing spaces at all; no trees, no flowers, no parks, no forest reserves and no pedestrian walkways: "In every corner, you will find one structure or another without adequate waste disposal system, unfriendly buildings facing and suffocating one another" (Act Three, scene one).

The statement indicates that the leaders negate efforts at climate change mitigation and that, in fact, they help to exacerbate the phenomenon in terms of both policy and action. The statement paints an ugly picture of an agonizing ecosystem. It shows that most of the actions of government officials tend to work against the environment and this seems to make the issue of climate change adaptation and mitigation something extremely difficult. In the text, the Chairman is used to illustrate a typical Nigerian politician. He sweeps the people off with grand speeches during electioneering campaign, but becomes a thorn in the flesh of the people on assumption of office. When the predicted flood eventually overruns the farmlands with its attendant devastation, Ifediba, one of the local farmers laments: "The fruit we saw from afar is no longer the same at a very close range. Is it not painful that it is the same jobless Edwin, who was almost on his knees begging for our votes just last year that has now refused to look into our plight? Is this the best way to reward us for the massive support we gave him?" (Act Three, scene two).

Performance and Reception

Schechner (1985, 120) argues that "An unproduced play is not a homunculus but a shard of an as yet unassembled whole". For him, performance is "twice-behaved behavior" and "Neither printing, scripting, nor writing shows actual behaviour as it is being behaved" (36). Drama, Schechner maintains, is "a loaded behavior multivocally broadcasting significances" (36). It is in performance and not in its literary form that drama reflects life. Wilson (2005, 10) cites Thornton Wilder to have stated that in dramatic presentation, we live in "perpetual present tense". Elaborating this, Eze (2011, 18) observes that "During dramatic performance the audience is given the impression that what is happening is taking place before his very eyes". The live nature of drama is, according to Wilson (2005, 16), "the single most important thing about its magic and longevity". Interpersonal contact which performance makes possible "ensures direct feedback between the actors and the audience, as well as between the spectators themselves" (Eze 2011, 33). This makes drama a lively and potent instrument of socialization. Enekwe (2007, 7) notes that during performance actors labour to "ensure that the messages are understood and internalized by the members of the community". Hodgson (1972, 26) avers that, "The power of the theatre to change the society is to implant images in the consciousness of the audience which represents the society".

Wake Up Everyone has elicited a number of performances. African Technology Policy Studies, Nairobi, Kenya, sponsored a twenty-minute production of a drama on climate change as part of its international conference on the subject in 2009. This was followed by a short performance of the same drama for the Climate Change and Adaptation Centre of the University of Nigeria, Nsukka in 2011. These two experiments subsequently led to research and publication of *Wake Up Everyone* as a full-blown play on climate change. Under this author's watch as the Head of Department of Theatre and Film Studies, University of Nigeria, Nsukka, the play was commissioned by the University Administration and was produced by his department as part of the events that marked the 44th Convocation Ceremony of the University, on 23 January 2015. The play has equally been staged by the Department of Theatre Arts, College of Education Ibadan and Bishop Ajayi Crowther University, Ibadan. Also, the Ministry of Environment, Climate Change Office, produced it in celebration of 2014 World Earth Day. Currently, the French Embassy in Nigeria is negotiating on the modalities of sponsoring its production in selected mega cities in the country.

The text has been well received in the country and has been extensively reviewed and discussed in major Nigerian newspapers and in the Ashden Directory on Environment and Performance. *The Punch Newspaper's* discussion of the play comes in form of an interview in which Akeem Lasis's "Dramatizing climate change" published on 6 March 2016 enables us to get an idea of the playwright's intention of using the text to puncture ignorance. The playwright, who appears shocked by the question some scientists in his university ask frequently, "What has drama got to do with climate change?" argues:

> Granted, the core science of climate change should stay within the domain of the natured sciences, but we cannot effectively disseminate information on climate change mitigation and adaptation without the skills, and expertise of dramatists and theatre artists in general. Theatre is one of the most effective strategies through which we can give those crucial climate change adaptation messages the depth of publication that

climate change scientists and even other organs of enlightenment can hardly give to the subject. (p. 45)

In *The Vanguard Newspaper* of 10 February 2011, Chidi Nwankwo in an article, "Greg Mbajiorgu's Wake up call: a theatrical wand against climate change", asserts that the play suggests that:

> All the scaffolding of the environment are constructs tied to the economy and power. The political elite control the means of production and resources of the state. Cupidity and avarice are instruments employed by the political elite to achieve immense financial aggrandizement. Conversely this process affects the economic lives of the workers who are left to luxuriate in abject poverty and destitution. (p. 20)

In his review of the play in *Daily Sun Newspaper* of 2 February 2012, entitled, "How Man destroys his own environment", Chidi Onah states that Greg Mbajiorgu's *Wake Up Everyone* is "… germane, ground breaking and expository. Mbajiorgu has done a noble work and set a pace in dramatization of the thorny issue of climatic change" (p. 36). Denja Abdulahi in an essay entitled, "Greg Mbajiorgu's play on climate change wins national award", in the *Saturday Sun Newspaper* of October 20, 2012, observes that: "*Wake Up Everyone*, a play by the University of Nigeria, Nsukka playwright, Greg Mbajiorgu, has won the Nigerian Universities Research and Development Fair (NUREDEF) which took place recently at the Federal University of Technology, Minna" (p. 41). Abdulahi further maintains that *Wake Up Everyone*, first experimented on stage in 2009 at the ATPS, Nairobi International Conference on Climate Change, "seems to have foreseen the flood ravaging the entire country today. Beyond warning about impending doom way ahead of time, this play inspires us to change the way we relate to our environment and dramatizes some of the measures rural farmers can take to survive the threats of climate change. *Wake Up Everyone* is celebrated for amplifying the viewpoints of rural dwellers on climate change adaptation issues and it is a theatrical demonstration of the complementaries between the sciences and the arts on this subject" (p. 41).

In the review of the performance of the play by the Department of Theatre Arts, College of Education Oyo, Hameed Olutoba Lawal, in an article entitled "Drama alerts govt. to the danger of flooding in a rainy season", published in the *Nigerian Tribune Newspaper* of May 2, 2013, argues that, "the opening scene of the play where Professor Aladinma, a dramatist and environmental activist visits the Chairman of the local government to alert him on the possibility of flood and the need for proactive measure… captures the attitude of a typical Nigerian politician who always turn deaf ears to wise counselling for communal development in preference to self-aggrandizement" (p. 26). He, however, notes that, "The resolve of the farmers in conjunction with professor's artistes to register their grievances, at the residence of the Chairman is a clarion call on Nigerian electorates to always make political office holders responsive and accountable" (p. 26). Henry Akubuiro, in "Greg Mbajiorgu: Garlands for foremost monodramatist", published in the *Daily Sun Newspaper* of May 22, 2015, pays a glowing tribute to the playwright and the play as well: "*Wake Up Everyone*, is Africa's first drama on climate change. And again it is the only play text in Nigeria that the National Universities Commission has awarded first prize at its inter-universities biennial competition" (p. 40). Finally, "Mbajiorgu: Our Scholars, Artists, Are Too Lazy to Innovate" is the title of an interview, which Artfolk of *The Guardian Newspaper* had with the playwright, published in the paper's edition of Sunday June 11, 2017. In this insightful discussion, the playwright suggests a practical way the Nigerian government can handle the problem of climate change:

> …. beyond the growing awareness on the subject in the academia, I sincerely wish that Nigeria, as a nation, would give the kind of attention it has given to the Boko Haram insurgency to the problem of climate change. I would suggest a Nigerian climate change mitigation and adaptation council, which will comprise experts from diverse field to proffer lasting solution on how we can tackle this planetary crisis (p. 28).

Apart from wide and favourable reception of the play in Nigeria, *Wake Up Everyone* elicited attention in the book *Landing Stages: Selection from the Ashden Directory*, where it is written about the play that "Wallace

Heim found truth and turbulence in the central character" (p. 122). This truth is that the play has a purpose: "to support impoverished farmers, to educate, to build resilience against the effects of climate change in rural Nigeria. The information on climate change is familiar, if uncomfortable. The role of experts in presenting knowledge to farmers is familiar, too, the belief and disbelief, the sometimes awkward juncture of different kinds of experience, the social power implicit in different kinds of knowledge." (ibid.)

Conclusion

This chapter has succinctly discussed the "argument" of the play *Wake Up Everyone* on the subject of climate change in Nigeria. The play contends emphatically that climate change adaptation and mitigation in the country face difficult challenges, emanating from ignorance, conservatism and mythopoeic assumptions. It maintains also that the major problem facing climate change project is government indifference, owing largely to a political culture that promotes self-aggrandizement and neglect of public causes. The play maintains that the idea of what to do is there, but that the all-deciding factor of money is in the hands of the politicians who are unwilling to make it available. The text laments that while the local dwellers seem to grow in knowledge about the subject due to enlightenment workshops like the ones Professor Aladinma provides in the play, and which could be a model for nationwide initiatives to raise people's awareness, government fails to create structures or take deliberate actions to stimulate the adaptive capacity of the people to deal with the threats of climate change. This pitiable situation, which frustrates genuine efforts, compels the Professor at the end of the play, when flooding devastated the people's farmlands, to tell Desmond: "I am tired, my bones are weak, the troubles of this land have wearied me" (Act Three, scene three). Above all, the play encapsulates the dirty game of oil politics in the land in Edwin Ochonkeya. Heim and Margolies lend credence to this by stating that, "In a single character the play conveys the immediate, turbulence, deceptive forces underlying oil production in Nigeria" (22). The text demonstrates clearly that oil extraction in the region has not been able to contribute to sustainable development, but on the contrary, has fostered terrible environmental hazards and debilitating poverty. Farmlands, air and water have been adversely polluted leading to poor agricultural yields and nigh extinction of aquatic life. This brings about quarrels between husbands and their wives as exemplified in Act three, scene two of the play, where Mazi Chinedum scolds his wife for demanding why she should prepare her soup again without fish. In fact, the text presents the common man as being mostly affected by climate change, while the government officials feed fat from compensation arising from oil spillage. This implies that the people are where they are because characters like local chief Edwin Ochonkeya benefit from the happenings. But the violent demonstration at the end of the play is an indication that this kind of situation endangers peace and stability not only on the family level but also in the society at large. However, the play articulates "functionally specific regimes to deal with an array of matters" (Young 1997, 2) that promote climate change. The theme song, entitled "Wake Up! Wake Up! Everyone", encapsulates all this, with its references to the need to build the world anew, not to burn down the forests, not to pollute the rivers, cut down the greenhouse gas emissions and the oil pollution, and "stop heating up the planet". The play is an unusual and powerful appeal to the public to take the environmental problem seriously and to inculcate its urgency in the Nigerian mind.

References

Abdulahi, D. (2012, October 20). Greg Mbajiorgu's play on climate change wins national award. *Saturday Sun Newspaper*, 41.
Achebe, C. (1983). *The trouble with Nigeria*. Enugu: Fourth Dimension Publishers.
Akubuiro, H. (2015, May 22). Greg Mbajiorgu: Garlands for foremost monodramatist. *Daily Sun Newspaper*, p. 40.
Akwanya, A. N. (2004). *Verbal structures* (2nd ed.). Enugu: Acena Publishers.
Artfolk. (2017, June 11). Mbajiorgu: Our scholars, artists are too lazy to innovate. *The Guardian Newspaper*, p. 28.

Awolowo, O., cited in Achebe, C. (1983). *The trouble with Nigeria*. Enugu: Fourth Dimension Publishers.
Azikiwe, N., cited in Achebe. (1983). *The trouble with Nigeria*. Enugu: Fourth Dimension Publishers.
Bassey, N. (2013). *Home run. Eco-INSTIGATOR: A publication of Health of Mother Earth Foundation*. Benin City: Health of Mother Earth Foundation.
Chidoka, O. (2015, January 22). *Rebuilding the Nigerian dream*. 44th Convocation Lecture of the University of Nigeria, Nsukka. https://guardian.ng/features/focus/rebuilding-the-nigerian-dream-mapping-the-building-blocks-2/. Accessed August 10, 2017.
Enekwe, O. O. (1990). Rich land, poor people: One view of underdevelopment. In Quintard Taylor (Ed.), *The making of the modern world: A history of the twentieth century*. Dubuque, Iowa: Kendall Hunt Publishing Company.
Enekwe, O. O. (2007). *Beyond entertainment: Reflections on drama and theatre*. Nsukka: University of Nigeria Press.
Eze, N. O. (2011). *The essence of drama*. Nsukka: Great AP Express Publishers.
Eze, N. O. (2013). Drama and the politics of climate change in Nigeria: A critical appraisal of Greg Mbajiorgu's *Wake Up Everyone*. *Rupkatha Journal on Interdisciplinary Studies in Humanities, 5*(1), 67–74.
Ezeliora, O. (2000). Aspects of African mythopoeia assumptions: Representation of the departed in Nigerian Poetry in English. *Okike: An African Journal of New Writing, 44*, 46–68.
Hodgson, J. (1972). *The uses of drama*. London: Eyre Metheun.
Kahn, R. (Ed.). (2010). *Critical pedagogy, ecoliteracy and planetary crisis*. New York: Peter Lang.
Kershaw, B. (2007). *Theatre ecology*. Cambridge: Cambridge University Press.
Lasis, A. (2016, March 6). Dramatizing climate change. *The Punch Newspaper*, p. 45.
Lenin, V., cited in Opata, D., & Ohaegbu, A. (Eds.). (2000). *Major themes in African literature*. Nsukka: Great AP Express Publishers.
Mbajiorgu, G. (2014). *Wake up everyone: A drama on climate change*. Abuja: Palace Press Ltd (originally 2011).
Nwankwo, C. (2011, February 10). Mbajiorgu's *Wake up Call*: A theatrical wand against climate change. *The Vanguard Newspaper*, p. 20.
Obi, C. (2008). A journey of a thousand miles: Researching conflicts in the Niger Delta. In *Africa in search of alternatives. Annual Report 2008* (pp. 12–14). Uppsala: Nordiska Afrikainstitutet.

Okolo, C. (1994). *Squandermania mentality: Reflections on Nigerian culture*. Nsukka: University Trust Publishers.

Okpoko, P. (2004). Environmental impact assessment and development decision-making in Nigeria: Lessons from the U.S. *African Journal of American Studies, 1*(1), 106.

Olutoba, H. (2013, May 2). Drama alerts govt to the danger of flooding in a rainy season. *The Nigerian Tribune Newspaper*, p. 26.

Onah, C. (2012, February 2). How man destroys own environment. *Daily Sun Newspaper*.

Opata, D. (2014). *"Blurb" of wake up everyone by Grey Mbajiorgu*. Abuja: Palace Press Limited.

Schechner, R. (1985). *Between theatre and anthropology*. Philadelphia: University of Pennsylvania Press.

Wilder, T., cited in Wilson E. (2005). *Theatre experience* (10th ed.). New York: McGraw-Hill.

Wilson, E. (2005). *Theatre experience* (10th ed.). New York: McGraw-Hill.

Young, O. R. (1997). Rights, rules, and resources in world affairs. In O. R. Young (Ed.), *Global governance: Drawing insights from the environmental experience* (pp. 1–24). Cambridge: MIT Press.

11

Future in Culture: Globalizing Environments in the Lowlands of Southern Ethiopia

Echi Christina Gabbert

Introduction

When looking at the peripheral lowlands of Southern Ethiopia today, we move into the midst of a dilemma that is symptomatic in our globalizing world. Territories used for centuries by agro-pastoralists were mapped out in the early twenty-first century by the government as investment zones and declared as unused, partly uninhabited or underused land (*terra nullius*) to be exploited for new agro-industrial goals. In rapidly changing circumstances, we observe divergent perspectives of different actors and groups of actors on land uses. Often these views seem irreconcilable.

The global neighbourhood approach, developed to analyse contact phenomena between hitherto unfamiliar actors (Gabbert 2014), tries to understand the different and often contradictory views and missions of the 'global neighbours'—local communities, policy makers, investors, NGOs,

E. C. Gabbert (✉)
Institute of Social and Cultural Anthropology,
Göttingen University, Göttingen, Germany
e-mail: christina.gabbert@uni-goettingen.de

human rights organizations, scientists et cetera—involved in the use of a particular territory, and wants to bring them together to discuss land use, mining, food security and agricultural production in the twenty-first century, in order to find points of convergence and constructive solutions.

This approach also entails dangers. Following Zygmunt Baumann, global neighbours can be divided into 'world makers' and 'world interpreters', or with reference to Huxley and Orwell into 'managers' and 'the managed' (Baumann 2000, 53f.). In global land policies, politicians and entrepreneurs mostly belong to 'world makers' and/or 'managers', local communities to 'the managed', social scientists such as anthropologists mostly to 'interpreters', and NGOs depending on their standpoint move between all categories. If we were able to articulate the different perspectives of the various parties, similar to an ethnographic study of different groups, positions could be represented as neat cases. But when describing the respective views of the investors, the administrators, the engineers, the locals and others, we might find ourselves in an arena of poly-perspectives which renders the approach useless because ultimately the world is made and managed by a few who are not necessarily interested in other points of view or the dynamics between them. Under such circumstances, positive features of cultural neighbourhood such as interest in one's neighbour, mutual respect and communication become irrelevant (Gabbert 2010). Ultimately in an imagined global neighbourhood which gives every viewpoint its own individual arena, neither dystopias nor utopias needed to be written anymore (Baumann 2000, 61) because we end up in a world of 'alternative facts', where every contested issue makes sense in each disconnected arena, and everyone can dwell in his or her version of reality. Such disconnection is not easily bridged, if at all.

One obvious reason for this is that people who decide or talk about land use are sitting in offices far away from those who know every tree, every plant and the seasonal variations of every water point in detail; planners construct, speculate and implement for others, while the expertise and experience of people who know most about the land are too often neglected. This asymmetry of interest and power is a decisive limiting factor and a constant challenge in multiparty settings such as land use and environmental politics (see also Parson and Zeckhauser

1995, 213). World makers, who are also world managers, install their plans and policies from a superior position which can ignore and crush different opinions and realities if they decide to do so.

What can we do when '[t]he usual analytic view of the world, which assumes symmetric interests, is not useful for most collective-good negotiations in the real world' (Parson and Zeckhauser 1995, 232)? What do we do when so-called stakeholder discussions are in fact exclusive settings of world makers and elites? In spite of all repetition, doubts and divisions, it is again time to oscillate between the extremes, to step back, write utopias and dystopias and by and within these dialectics listen to the emerging tenor of global neighbours, their grounds, their possibilities and abysses. Rather than judging one or the other, it is time to 'bring ideas into life that matter'.

In search for a tenor of dissonating thoughts, in the following I first describe a global dystopian scenario, then a utopian scenario of the Ethiopian context, and finally present an ethnographic example from Southern Ethiopia as basis for discussion.

Dystopian Scenario

Dystopia: an imaginary place or condition in which everything is as bad as possible.[1]

World Makers and World Interpreters

> What they [Huxley's and Orwell's dystopias] shared, was the foreboding of *a tightly controlled world*; […] of a world split into managers and the managed, designers and the followers of designs - with the first keeping the designs close to their chests and the second neither willing to nor capable of prying into the scripts and grasping the sense of it all; of a world which made an alternative to itself all but unimaginable. (Baumann 2000, 53f.)

At the beginning of the twenty-first century, negative consequences of the Anthropocene become very apparent. Environmental degradation and exploitation are not under control, and environmental disasters such as oil spills and nuclear and chemical pollution constantly take their toll on the health and well-being of humans, flora and fauna. Climate change and loss of biodiversity pose a serious threat to humanity (Chivian and Bernstein 2008).[2] In 2017, while people of all regions in the world suffer from the devastating effects from extreme floodings (e.g. Africa, Asia, Latin America, USA) and droughts (Eastern Africa), we have entered a stage of climate change denial that tries to show that decades of research is not only unwanted but is actively targeted in a country as 'developed' as the USA, a country where a perpetuation of overproduction and food insecurity live side by side (Dunlap 2013).[3]

While forms of small- and medium-scale agriculture, often based on agricultural heritage systems and agro-ecology, are recognized as well suited to feed communities on sustainable grounds (AFSA 2016; Altieri 2009; FAO 2014; IFAD 2013; Koohafkan et al. 2012; TWN and SOCLA 2015; UN 2013)[4] and the resilience and flexibility of agro-pastoral systems with their socio-ecological expertise and risk management have been described in an ever-growing number of works (e.g. Abbink et al. 2014; Catley et al. 2013; Krätli 2015; Lane 1998; Little et al. 2001; Schlee 2013a; Scoones 2004; Zinsstag et al. 2016), mono-crop models, industrial agriculture and meat production are heavily promoted and implemented all over the world. Although industrial meat and dairy production especially in the western countries 'are huge contributors to global climate change' (GRAIN 2017), especially in so-called untapped regions of the world, large-scale projects are replacing and destroying time-tested agricultural and agro-pastoral knowledge systems and ways of living. Many fine works on ecology and food production provide a warning body of science that stands in opposition to the path taken by politicians and agro-industries, and there is a tendency to politicize and discredit scientific reports and local knowledge when they do not fit into the presently dominant ideology. Findings are used selectively to support policies rather than for critical engagement and improvement of the challenges at hand. High modernism is legitimized not only by constructing its opposite: the underdeveloped, vulnerable and/or backward peasant or pastoralist living on so-called wastelands (see also Scott 1998, 330), but also by silencing its critics through words and deeds.

In these divergent settings, world makers and world interpreters, for example scientists and politicians, do not come together in a way that facilitates the mutual measures that would secure the health of our planet and the basis for food security for future generations. More importantly, small-scale and family farmers who provided the backbone of food security are more than often left out of the planning procedures and are among the ones paying a high price for the planning from afar.[5] Instead, sustainability, a catchword for agricultural policies, is being led *ad absurdum* as the destruction and exploitation of finite global resources is being pursued at a pace that eludes sustainability but follows business patterns of enrichment, steered by a small number of international corporations and elites who control how and what to produce, and at what price without considering the well-being of coming generations.

Negative effects of global fossil fuel-dependent industrialized agriculture are seen and felt globally. Monopolies on seeds are one example of tightened dependencies on and restricted agency of farmers on commodified markets of agriculture. Mono-crops need external inputs and tight pest control that have led to a 'pesticide treadmill' in many regions (TWN and SOCLA 2015). Soils are exhausted, and in many places, nitrate and phosphor levels in water and soils have become a serious problem as a result of overuse of fertilizers. This threatens both biodiversity and people's health (Magdoff et al. 2000).[6] In spite of this knowledge, environmental costs and health costs are mostly excluded from calculations of revenues for industrial agriculture.[7]

In 2007/2008, the food price crisis sparked off a new scramble for land all over the world. The spectre of food insecurity was used as an argument to speculate with land as commodity worldwide in spite of overproduction, waste and loss of food in the regions where most speculators live (White et al. 2012).[8] Instead of designing balanced global food distribution and preparing for changing farming conditions in times of climate change by seriously working on the malfunction of rigid systems of agricultural overproduction and food waste especially in western countries, many food producers, and consequently many land speculators, used the opportunity to secure more land to exploit in so-called developing countries (see also White et al. 2012, 627). Parallel developmental efforts, for example in African countries, to improve national economies, agricultural production and/or food security were a perfect match for the emergent speculation on land. In Ethiopia, for instance, large tracts of land were secured for sell-out leases[9] and industrial production was hardly hampered by regulations regarding the use of agro-chemicals, especially compared to rules and regulations in the investors' home countries. On the contrary, fertilizers and pesticides were promoted among rural communities, often without providing sufficient information about their possible negative effects on soil and people. In this way, millions of hectares of land were leased to investors to develop the peripheral regions of Ethiopia (Dessalegn 2011, 2014).

In 2017, within a decade Southern Ethiopia as well as other regions in Ethiopia has been made available for national and global market schemes. The fast-track development of regions and people of the country labelled as 'backward' is central to the program of a new developmental Ethiopian state, which aims at integrating the Ethiopian peripheries into ambitious schemes within the framework of the Ethiopian 'Growth and Transformation Plans' (MOFED, FDRE 2010; National Planning Commission, FDRE 2016) and the UN Millennium Development Goals. The dynamics and background of these development efforts also regarding climate change have been described and analysed by scholars from different fields (e.g. Abbink et al. 2014; Abbink 2011; Abebe and Solomon 2013; Asnake and Fana 2012; Avery 2012; Borras et al. 2012; Buffavand 2016; Cotula 2012; Dereje 2013; Dessalegn 2011, 2014; Fana 2016; Fratkin 2014; Gabbert 2014; Galaty 2013; Girke 2013; Kamski 2016; Lavers 2012a, b; Makki 2012, 2014; Meckelburg 2014; Scoones 2004; Strecker 2014; Tewolde and Fana 2014; Turton 2011, 2015; Wagstaff 2015; Yidnekachew 2015). The environmental and social impacts in the affected areas have been criticized by Human Rights Organisations (e.g. HRW 2012; Oakland Institute 2011, 2013). A recurrent observation is the negative effect of top-down development approaches that forego the potential benefits of seriously integrating and respecting local knowledge, experience and sociocultural realities. Administrative efforts to communicate with local populations came too late in the procedures and initially left out people who were drastically affected by changes to their livelihoods. Indeed, villagization schemes to make way for commercial farms have led to hardship for many agro-pastoralists who not only lost their basis for subsistence economy but their aspirations for a future of their own choice, while promises of a brighter future and trickle-down benefits have not manifested for most people. 'The encroachment on their rangeland follows a familiar path seen many times over: closure, livelihood subversion, impoverishment, and decline' (Abbink 2011, 517). This results not only in economic insecurity and hitherto unknown dependencies, it also creates a feeling of loss of a very existence within an environment that was inextricably connected to people's livelihoods, identities and their being in the world.

As of 2015 in Southern Ethiopia, agro-pastoralists around the large sugar cane plantations (which started in 2012) along the Omo River tell of severe shortages of water and food and hitherto unknown dependence on external sources and help, as the agro-pastoral practices, as well as the trade and exchange networks of the whole area, have been disrupted due to the plantation schemes.[10] Flood-retreat cultivation at the banks of the Omo River is not possible since the water has been filling the hydroelectric Gibe III Dam further north. Controlled flooding has been insufficient and badly coordinated. Animal husbandry with drastically limited access to water and of dry season pastures is seriously impeded. Traditional agricultural activity along the river has been shattered, and 'compensation' plots away from the river are dependent on the services provided by the plantation with machinery and irrigation. Some people have not received compensation plots; others have not been able to cultivate their fields because of delays in the services. Food shortages due to failed harvests cannot be compensated anymore by time-tested flexible traditional practices, nor by upscaling of animal husbandry. Monetary gains for a small number of local workers on the farms are not compensating this loss of subsistence economy for entire families and communities. In this scenario, the narrative of the 'food insecure' or 'poor' which was used to legitimize the changing use of people's land into agro-industrial territory in the first place has come full circle. People who did not perceive themselves as poor before are now deprived, not only of the means needed to practice subsistence agro-pastoralism but also of their ways of being that encompassed much more than the immediate material use of land. Identity, cultural integrity, ways to care for the land that connects the ancestors with future generations, a being in the world connected to paths and waters and shade of the trees, to animals spirits and humans, were obliterated by planners who decided for, instead of with, people about what is good for them (Buffavand 2016; see also Lydall 2010, and Abbink, Chapter 6, this volume).

From a distant and, say, technocratic standpoint, the sacrifice of livelihoods for fast-track development based on export-oriented production of cash crops is an unavoidable, if painful, consequence for a few for the benefit of the whole: the overall industrialization of a country. The calculated 'cash nexus' from foreign earnings reigns over long-term social, cultural and ecological considerations, especially for planners who are far away from the negative effects of their planning. But first, the overall benefit of the agro-industrial projects has not yet been realized, with several investors abandoning their projects for different reasons.[11] Others have not met the economic and developmental expectations (Kamski 2016), and still others have met more local resistance than expected. Second, the loss of cultural identity and economic independence which was bound to land and livelihoods, goes almost unmentioned despite its devastating effects on agro-pastoral communities and on the environment, ignoring the grim realities of such dispossession (Turton 2015). Human rights organisations' reports on the socio-ecological effects were dismissed by the Ethiopian government (HRW 2012); to boot, several scholars found themselves in difficult fieldwork situations, after having been labelled as 'development spoilers'.

Between these hardened fronts, the sensitivity of personal fates and a serious integration of sociocultural and ecological knowledge do not find a proper space. The personal suffering of people experiencing the decline of their livelihoods is hardly describable. Additionally, the resentment and conflicts that this dynamic triggers negatively affect economic success and national security ('tenure risk').[12] Tensions between farm workers and local populations are resulting in violent incidents in different regions of Ethiopia.[13] In Oromiya, people's discontent about not having being sufficiently involved in land use and policy planning led to violent protests, beginning in 2015. In 2016, eruptions of violence, which also led to attacks on farms, were countered by the declaration of a State of Emergency in Ethiopia in October, leaving Ethiopia divided along different lines of mutual resentment that destabilize the country.

As a matter of fact, to pave way for development, the government took the risk to quickly 'melt the solids' (Baumann 2000, see below), such as cultural values and socio-economic relations that many Ethiopians cherished and identified with. Not only the sacrifice of livelihoods, family farming, socio-ecological expertise and security were taken into account but more than that, the management and control over the peripheries became a prominent feature of capital accumulation (Fana 2016). Let me end this dystopian view here with a passage from Zygmunt Baumann (2000, 4):

> 'Melting the solids' meant first and foremost shedding the 'irrelevant' obligations standing in the way of rational calculation of effects; as Max Weber put it, liberating business enterprise from the shackles of the family-household duties and from the dense tissue of ethical obligations; or, as Thomas Carlyle would have it, leaving solely the 'cash nexus' of the many bonds underlying human mutuality and mutual responsibilities. By the same token, that kind of 'melting the solids' left the whole complex network of social relations unstuck - bare, unprotected, unarmed and exposed, impotent to resist the business-inspired rules of action and business-shaped criteria of rationality, let alone to compete with them effectively.

Utopian Scenario

An imagined or hypothetical place, system or state of existence in which everything is perfect.[14]

Hand in Hand

> In this regard, the key resource that we have to primarily rely on to sustain the on-going development effort is our people. (National Planning Commission, FDRE 2016, 222)

> Ethiopia has the lowest per capita consumption rate of fossil fuels. It has almost no contribution to the greenhouse gas emissions that are the cause of adverse climate change impacts. On the other hand, in Ethiopia there are indigenous knowledge, practices and systems which reduce biogas

emissions. In the agriculture sector farmers use biologically based inputs like cow dung and compost as fertilizers. Ethiopia's contribution to creating a stable and more beneficial climate is very clear. (MOFED, FDRE 2010, 120)

Ethiopia is regarded as the home of humanity.[15] It is the only African country that does not have a legacy of external colonization.[16] The country has a great diversity of livelihoods and cultural knowledge. With its rich history and natural and cultural potential, Ethiopia engaged in fruitful exchanges in fields of diplomacy, international relations, agriculture and livestock, arts and sports as well as scholarship in all fields. Culturally diverse regions and populations provide unique resources of political, philosophical and socio-ecological knowledge, with century-tested agricultural and agro-pastoral production techniques still in active use. In the following, I hypothetically spell out the potential for more sustainable, equitable and adaptive land use and transformation in Ethiopia.

Peripheral regions, at some point mistaken as 'empty lands' or 'unused territories', and their populations, at some point mistaken as 'backwardish', provide invaluable knowledge that has been lost in many other regions of the world. The merging of traditionally innovative expertise, such as agro-pastoralism with modernization, provides a unique opportunity to develop urgently needed resilient production techniques in times of climate change and upcoming global environmental hardship. Socio-ecological practices and solutions that are being recreated and reconstructed in over-industrialized regions of the world with great effort (such as agro-ecology, permaculture, conservation agriculture) are still fully functioning in many regions of Ethiopia, for example in Southern Ethiopia and the Gamo highlands. In particular, agro-pastoralist communities have developed techniques and measures to survive under harsh conditions and seasonal variation in territories that would be unsuitable for intensive cultivation. As climate stability will become a condition of the past in more and more areas of the world, the flexibility of these systems has the potential to provide the world with much needed knowledge, skills and practices to react to climatic fluctuations with a combination of seed variety, cultivation, and watering techniques and other adaptation methods (Krätli 2015; Manzano 2015; Gil-Romera et al. 2011). Thus, the innovative character, skills of improvisation in changing environmental circumstances (see also Hallam and Ingold 2007) and non-equilibrium approaches of agro-pastoralist knowledge, in other words living and dealing with uncertainty (see also Scoones 2004), provide guiding insights for agricultural, ecological, sociological, political and philosophical aspects not only of development policies but also of well-being in a more holistic sense.

Agricultural enhancement measures are based on existing culturally diverse land-use approaches to meet the challenges of food security, while industrialization projects are carefully weighed against local land uses and customary practices to find the best solutions for particular regions with the aim to avoid destruction of irreplaceable systems of biodiversity and traditional knowledge.[17] If, after careful inspection of all alternatives, new farms or other enterprises are established, this is done with full inclusion of the local communities in the process, not only providing transparency of the investment schemes but also committing special attention to existing knowledge and ensuring environmental and social responsibility. The time and money this takes are part of the investment schemes, as realized, for example, in parts of the operations of an oil firm in Southern Ethiopia that provided quality trainings to local workers instead of importing specialists from other regions. The same firm also put much effort into establishing neighbourhood relations on the basis of cultural respect, relations that were developed and cultivated throughout the operation, leading to a code of ethics adhered to by the external national and international staff.[18]

With due effort, friendly neighbourhood relations help realize common goals as well as face diverging interests peacefully. The limits involved are acknowledged with special attention to customary land-use patterns and ethnic identities (see also Schlee 2013b). Compensation measures are taken seriously, and emerging benefits through cooperation between new neighbours are realized and practiced. This has partly happened in 2017 when revenues of a Sugar Plantation were provided to feed livestock in communities in the Southern Nations, Nationalities, and People's Region (SNNPRS) that had lost their pastures due to a devastating drought (personal communication with several people in South Omo).

Economic visions for the future are studied thoroughly in emerging centres of excellence at Ethiopian Universities and in the field and are realized in cooperation with local communities. The potential of pastoralists to produce high-quality livestock products rather than cheap mass production for the market is acknowledged (GRAIN 2017). In this cooperative landscape of expertise, the exchanges between practitioners, engineers and scientists, farmers, agro-pastoralists and universities drive Ethiopian innovation that in turn enriches global knowledge systems, redefining sustainability in positive, grounded and realistic terms.

Technology departments excel in developing the climate-friendly politics of the Ethiopian government with Ethiopia's enormous potential for renewable energies that provide the population and the market with clean energy. On- and off-grid solutions, such as solar-, wind- and geothermal energy, combine the natural potential of people's habitat with sovereign solutions. Western countries live up to their environmental global debt (such as harmful emissions) by openly providing the latest knowledge on renewable energies to be further developed in Ethiopian technology centres.[19] Economic and social science departments redefine corporate social responsibility (CSR) schemes for investments on land in close cooperation with local people and circumstances, thus providing realistic grounds for corrections of western models of development and CSR. These models gain worldwide recognition as examples for equitable development. The cooperative and equitable approach is able to tackle and decrease tenure risk and increase identification with common national goals (see also Gabbert 2014; Strecker 2014). Schools that are respecting local livelihoods and knowledge, including functioning mobile schools, contribute to a fruitful combination of school education and customary education (see also Krätli and Dyer 2009).

Knowledge about medicinal plants and biodiversity is combined with modern medicine and environmental protection measures to improve national health standards for both humans and animals. Hitherto unknown medication is studied and refined into Ethiopian products that contribute to the global demand for natural medicine.

Combining these fields and resources, Ethiopian policies enhance thoughtful concerted national and global efforts while building on local expertise to develop different branches of innovative and mutually respectful generated knowledge and policies that support national production of food, medicine and energy. Thus, Ethiopia sets a unique example for international acclaimed findings in all fields, providing invaluable keys for food security, national security and peace, citizens' well-being, global health, climate-friendly policies and technologies as well as economic self-sufficiency, realizing the vision to generate African solutions to global challenges in a long overdue South–North exchange. Let me end the utopian view with a statement from an Arbore man from Southern Ethiopia:

> If things are developed peacefully and respectfully, everyone can learn and listen and change and open their minds. But if we are treated as inferiors, where is understanding, where is respect, where is the peace of mind that opens minds and hearts? If things are told patiently, they will be listened to. If you are forced to eat, will you feel well after the meal? You will not. If you eat in a friendly atmosphere, the meal will nourish you. The same happens with conversation and knowledge. Let us nourish each other. (Interview with Arbore man, September 2016)

On the Ground—Arbore in Southern Ethiopia

> My uncle's white cow
> The land where the grass is sweet smelling
> This is where I take her
> (from an Arbore herding song)

Knowledge of the environment is the best basis for its proper use and development. But how many people still have such good knowledge of their environment that '[consists] in the skills, sensitivities and orientations that have developed through long experience of conducting one's life in a particular environment'? (Ingold 2000, 25).

In Ethiopia, many people combine animal husbandry with agriculture in harsh environments where non-equilibrium ecosystems are the norm. The constant challenges of seasonal unpredictability are part of subsistence systems that are extremely flexible and adaptive. Among those agro-pastoralists, the Arbore of Southern Ethiopia, a group of

9000 people, practise a mixed subsistence economy based on agriculture and animal husbandry. It would be insufficient to describe the Arbore way of life in purely economic terms. Both in agriculture and in animal husbandry, the relation of subsistence economy to social, ecological and ritual understanding is inextricably bound to the Arbore's understanding of being part of the land and the world together with their animals. The use of pastures and water is coordinated by the community. Although families tend to use fields for several years, customary landuse rights are never individual nor are they 'forever' (see also Harvey 1996, 223). Land is perceived as Arbore territory but the Arbore perceive themselves as bound to the country rather than owning it. As all things are of the Sky God (Waq) land and people are as well and therefore have to be treated with consideration and respect.

Agriculture

Combining flood-retreat, slash-and-burn, and rain-fed cultivation with irrigation, shifting cultivation and a seed pool of more than a hundred varieties of sorghum, maize and, to a lesser degree, beans, pumpkin and other vegetables, the Arbore produce their major staple foods under difficult climatic conditions. Dairy products, meat and animal blood form another source of nutrients, with leaves, berries and fish adding to a variable diet. Storage of grains, dried meat and dried fish prepares for times of shortage.[20] The climate in Arbore is semi-arid, with high seasonal variations between rainy and dry seasons. The Woyto River that crosses the Arbore territory only temporarily carries water and has changed its course several times over the past decades.

To secure harvests and pasture access for animals under these all but favourable circumstances for agriculture, the Arbore combine a refined and flexible set of agro-pastoral techniques which integrates all community members and their social circumstances in the distribution of land and production processes. Their elaborate and adjustable practices have enabled Arbore to become a 'bread basket' of the wider region in spite of the difficult climatic and geographical circumstances.

The distribution of arable plots is overseen by men appointed to the task (*mura*) and is a complex matter of social and ritual importance based on the Arbore generation-set organization (Ayalew 1995, 85ff.; Miyawaki 1996; Gossa 1999, 166ff.). The distribution of the most fertile floodplains, demarcated with stakes, to the community takes changing seasonal variation in rains and floodings as well as social circumstances into consideration and is accompanied by rituals to bless and protect land and people. Plots that are easier to cultivate are given to senior elders, especially elderly single women, whereas plots that need more work are given to larger families with younger adults. In case of less inundated land (due to reduced flooding in a given season), the former distribution to families is adjusted to give everyone opportunities to secure their food. Dates of sowing and harvesting are communally synchronized, so that pests like birds and other animals evenly distribute over all fields. This lowers the possibility of total harvest loss for single plots and families. After the fields have been harvested, they are opened for herders to feed the cattle with the remains of the plants and the temporary shelters, all made of plant material. In addition to flood-retreat cultivation on the most fertile fields, which are divided among the community as described above, individuals can make areas with dense vegetation arable with slash-and-burn cultivation. These fields (*luuch*) either depend on rainfall or can be irrigated if close enough to the river. The Arbore also dig irrigation channels depending on the position of temporary river flooding. If the water for irrigation is scarce, users of neighbouring fields coordinate the opening of the channel system so that every plot receives an even amount of water. Rains as well as the floodings of the river are expected in certain timeframes—of large major (March–May) and small rains (October/November)—but remain unpredictable, with varying water availability. Specific sorghum varieties, for example drought-resistant and quick-growing species versus slow-growing and moisture-requiring species, are chosen according to seasonal, nutritional factors and availability of family members to protect the ripening crop. Bitter species do not need to be protected from birds, as much as sweet species, and are preferably cultivated by families with fewer or younger members. Each family uses a variety of species on each field according to their family needs and preferences, and to reduce the risk of total crop failure.[21] If harvests fail in spite of this buffered system, the family, clan and settlement provide a final safety network to supply a family with staple food. The ultimate insurance factor for every family is the herds. In particular, in times of famine, milk and blood from animals provide the main nutrients, and in times of drought, animals are traded for grain from neighbouring groups.[22]

Animal Husbandry

In spite of their extensive agricultural knowledge and production, the Arbore's self-reference as 'people of cattle' (*modo oot*) is most salient and runs through all spheres of Arbore life, connecting individuals, families, lineages, clans and even ethnic groups. Cattle and also goats and sheep are important means through which kinship, friendship and bond friendship are enacted and celebrated in rituals and on a daily basis. Each day in Arbore is framed by *bobba*—the moment when the cattle leave the kraal in the morning—and *galshum*—the moment when they return in the evening. Providing good water and pasture for the family's livestock is not only a daily exercise but part of a person's identification as pastoralist and requires exact knowledge about weather, climate and environment also reaching into the wider cultural neighbourhood between pastoralist groups. Domestic herds and camp herds are managed by Arbore of all ages, boys and girls, men and women. Cattle determine much of the Arbore's material, oral and musical culture and symbolism. Cattle connect people within and between ethnic groups, and every important step in a person's life is accompanied by transactions of cattle, goats or sheep. Cattle are means to establish social networks with an 'insurance factor' (Schlee 1989, 402ff.). For a child's first hair-shaving ceremony, for marriage and at a person's death, family members and friends contribute cattle, and social ties are constantly re-evaluated according to these transactions. The land for cultivation is blessed with sacrificed animals. Items made of these animals, such as the hides of oxen slaughtered at the wedding ceremony, carry practical and symbolical value.

Normally, the first children of a married couple are conceived and born on such a cowhide. Some ritually important adornments, such as the mother's belt (*mach*), must only be made from the hides of healthy and properly sacrificed animals, to transfer the strength of the animal and thus increase the object's mystical powers. The patterns on that belt, which are made of cowries, are divided by a symbolic cattle path. Worn and ragged cowhides can be used to carry away old bones and baby droppings outside the settlement. Cow's milk is of nutritional and symbolic importance. In particular when the harvests fail, children can survive on milk. The most frequently used liquids for blessings are a mixture of water and milk (*err*) and butter from cow's milk. The handing over of the peritoneal fat (*moor*) of slaughtered livestock, a part of every major ceremony, is one of the most prominent symbols for peaceful relations. Cattle terminology is pervasive in the Arbore language, especially in colour classification, and many Arbore names are derived from cattle's features, colours and patterns. The naming of a child is sealed by placing the umbilical cord on the back of a cow. In their youth, boys can add the name of their favourite animal to their given names. The pastoralist ideal also determines the socialization and education of children. All Arbore girls and boys are taught to care for the family's livestock. To become a good and responsible person is to become a good and responsible herder. Most of the Arbore who attended school and are now employed in administration, by NGOs or other enterprises, still share pastoralist values. I conclude this passage about Arbore pastoralism with the words of a young Arbore, whom I asked about the difference between cattle and money:

> Our wealth is alive, out there, visible and beautiful. You can respond to it with your heart and with good energy. Money is pale and seems like an illusion; when you hold it, it does not give you good energies; when you look at it, it does not cause your heart to rejoice. When you pay with money, the transaction is over. When you pay with cattle, the connection is just beginning. Cattle open as many ways between people as they have hair on their hide. Every transaction of a cow has to be discussed thoroughly. Every matter of cattle brings together many people and opinions and has to be considered carefully.

Discussion and Conclusions

> If you listen, peace will open up.
> (Arbore man, interview, September 2016).

Development efforts in any country are first and foremost positive visions to improve people's livelihoods. To respect and support these visions is the basis for constructive and peaceful discussion. Likewise, to respect and understand people's views on land, especially those that are most familiar with their land and most affected by changing land uses, is not an optional act of benevolence but an essential part of the opportunity to create the best possible solutions to pressing challenges to food security worldwide. Even if we have to admit that people in global food arenas play with different means and aspirations and some people continue to enrich themselves at the costs of others, this makes the encounter less pleasant and cooperative but not less important.

To write about people's views on land use in dystopian and utopian terms is an attempt to place seemingly opposed visions that have formed and hardened in a new relation that inspires us to face and question the challenges at stake. Neither vision works alone, but before judging one or the other vision may be as hostile to economic growth and eco-centred fantasy or vice versa, one can step back and accept veritable perspectives and aspirations with different means and backgrounds all hopeful for a future that lies in people's hands. The question about whose hands are meant and meaningful remains. In the dystopian view, the decisions lie in the hand of the managers for the managed; in the utopian version, people work hand in hand; and in the practical example, the hands that work the land secure the food for the families, the community and beyond. The outcome of this exercise with reference to anthropology will be discussed in the remainder of this chapter.

'By default, and also for independent reasons, economics has become the science of the future', whereas the 'cultural actor is a person of and from the past' (Appadurai 2013, 180). This division, as Appadurai contends, is a crippling one, both for people and disciplines and for gaining meaningful insights for future-making. The role of social sciences and of anthropology is to bring back the cultural into the discussion, not as academic exercise, not as magic formula, not as romantic reminiscence but as realistic and indispensable element of innovation and change. 'Culture, in other words, is not a mere object, acted upon by nature, but a subject which constitutes or gives meaning to nature' (Turton 1985, 331). Cultural knowledge transcends the material use of resources and encompasses considerations of land use, conflict, commonality, environment as well as change (see also Baviskar 2008). Cultural knowledge can make economic aspirations and practical solution-finding more, not less realistic. Cultural knowledge and precision 'distinguishes the skilled practitioner from the novice' (Hallam and Ingold 2007, 12). In short, cultural knowledge is indispensable for sustainable food production.

In the Anthropocene, we have reached a point where the finiteness of our world is our everyday reality and it will be even more so for our children. Not addressing this responsibility with all available resources is not excusable. Of course, when criticizing economic aspirations of so-called developing countries, one has to be aware of the inherent danger and arrogance of that endeavour. Much damage done to our environment is created in fossil fuel-based industries in western countries with a culture of consumerism that uses up more and more resources of the world. This regional wealth comes at the expense of the entire world, but is especially damaging for so-called developing countries. To ask of those countries not to add to this global mess while having to improve living conditions for a growing number of their citizens indeed is difficult, both politically and ethically.

Therefore, I want to stress what anthropologists and other scientists do not stand for: anthropologists do not want to preserve culture, because culture per definition is always dynamic and anthropologists are the last to claim otherwise. Yet anthropologists respect and integrate culture because they have studied practices and knowledge that offer impressive coping mechanisms for non-equilibrium circumstances. These are and will be of increasing importance in times of climate change not only in local detail but also in global comparison, such as the fine-tuned and flexible agro-pastoral system of the Arbore. This does not mean that all is well in this or other agro-pastoralist settings, or that solutions can or should be simply transferred, but in a globalizing world, regional knowledge is a hidden and under or even devalued asset that—if recognized—enables us to question not only our production practices and enrich necessary improvements, but also challenge our value systems. Such scrutinizing includes wrongly simplifying labels such as vulnerability and poverty or concepts like 'development' and sustainability. Ethiopia with its federal organization had prepared unique political grounds for the meaningful integration and potential of particular cultures, values and knowledge, but its policies are currently eradicating many local particularities with planning that too often decides *for* people, not *with* people, about their very future. To justify this, traditional knowledge systems have been stigmatized as 'backward' so as to reorganize or sideline them from above. Traditional subsistence livelihoods that contribute to national food security have typically been labelled as 'vulnerable'. Without any need to romanticize, there is a need to stress that it is not tradition that hinders progress. As the German philosopher Marquard (2003) contends, future needs its past, its heritage and its origin (*Herkunft*). Indeed, there is a need to scrutinize simple assumptions and ask questions anew. What is wealth, what is well-being, what is progress and what is a future to be aspired to in a finite world? What if local knowledge and practices are accepted as keys to develop better solutions, especially when addressing failures that are tinkered upon the western world?

In the utopian section of this chapter, I sketch possible improvements developed by people with different backgrounds from pastoralists to farmers to scientists feeding each other's knowledge while working hand in hand, merging fields of knowledge. With its historical depth, cultural diversity and innovative potential, Ethiopia has many means to combine global technological knowledge with local expertise about land use and ecosystems. To realize potential of knowledge long lost in western countries is an asset beyond the mere tapping of natural resources. To take care of resources without shortcuts and hasty calculations is crucial in a finite world. Here, the knowledge and values inherent in traditional subsistence systems can lead the way to more innovative socio-ecological and political economic alternatives than any blueprint planning.

Improvements such as stated in the Millennium Development Goals are much needed, but should not lead to people being driven into scenarios that hinder them in making their own choices about a good future. Choices can move from working on a farm as labourer to be a good pastoralist or many other possibilities. The peaceful coexistence of different livelihood options needs careful and cooperative planning. To hurry slowly (*festina lente*) is not only a western concept but can be found in many cultures in the world, also in Arbore. '*Nungu, nungu*' (slowly, slowly), the elders tell the young ones before they rush into a decision, an argument, a fight or trouble. To do things slowly is grounded in the maturity to carefully consider the consequences of each decision and to reach the best possible outcome for everyone involved. Whereas remarkable progress has been made in facilitating economic development and better infrastructure, many fast-track investments in Ethiopia have failed in less than a decade, for different reasons. With less haste, many of those failures could have been prevented and with it the financial loss and socio-ecological turmoil and hardship. With less haste, the Ethiopian peripheries would never have been declared as 'empty spaces' at the first place, or more precisely, spaces unrecognized in their unique potential in the minds of the planners (after Baumann 2000, 103).

Therefore, the third part of this chapter that sketched the elaborate agro-pastoral system of the Arbore is not another pledge to seriously integrate local knowledge into the planning. Local knowledge is not another tool to be used for development planners. The conclusion goes further: planning can only happen in partnership, informed and scrutinized most importantly by people who know what it means to care for the land in all its complexity, not for mere benefit but for survival, not for professional merits or greedy speculation but for securing real harvests for real people, not for individuals but for a whole community, not for today or tomorrow but for generations to come. Arbore mark the land to be distributed with stakes every season again to decide about the best possible harvest outcome for every single person in the community in an ever-changing environment. Culturally grounded skills in improvisation and innovation in changing and unpredictable environmental circumstances cannot be underestimated, especially in times of climate change. Serious planners will acknowledge that local populations, small- and medium-scale farmers and agro-pastoralists do not need to be developed on a drawing board far away from the realities of their environment. To call them partners, who might serve as another element in the planners' toolbox would also not suffice. To fully recognize that *they are* indeed the most important stakeholders, fellow planners and world makers, whose knowledge is constantly developed in real life, would open immense sources of creativity and innovation and would place them as necessary specialists in the centre of designing, guarding and managing the use of land. People who know their land need to be central in addressing and solving present challenges about climate and food and they need to speak first and not last, as visionaries and not as objects of the past, bringing ideas into life for the future of the lands and the world. Perhaps invent them. All for the better.[23]

Acknowledgements I thank the Max-Planck-Institute for Social Anthropology in Halle/Saale and the Institute of Social and Cultural Anthropology, University of Goettingen, for funding and supporting my research. Jean Lydall, Lucie Buffavand, Felix Girke and Ivo Strecker have commented on earlier drafts of this chapter.

Notes

1. 'dystopia, n.'. *OED Online*, June 2017, Oxford University Press (http://www.oed.com/view/Entry/58909?redirectedFrom=dystopia, accessed August 29, 2017).
2. The prize-winning volume *Sustaining life: How human health depends on biodiversity* (Chivian and Bernstein 2008) is an outstanding illustrated contribution to the discussion by more than 100 leading scientists.
3. National statistics state that 42.2 million Americans lived in 'food insecure' households in 2015 (Coleman-Jensen et al. 2016, 6).
4. For a critique of the misuse of agro-ecology by agro-industries, see Giraldo and Rosset (2017).
5. 'Fifty to seventy-five per cent of the world's food is produced by small farmers, even though they only control 25-30 per cent of the land, and use 30 per cent of the water and 20 per cent of the fossil fuels used in agriculture' (TWN and SOCLA 2015, 38).

6. For Germany, see reports of the Federal Environment Ministry BMUB and the Federal Ministry of Food and Agriculture BMUB and BMEL (2017).
7. '[…] the current cost of food is actually higher when we account for greenhouse gas emissions, water contamination, loss of biodiversity, soil losses, public health impacts and other externalities. In the UK, the price tag of industrial agriculture's externalities is about £205 per hectare' (TWN and SOCLA 2015, 4).
8. According to FAO, almost one-third of food produced for human consumption—approximately 1.3 billion tonnes per year—is either lost or wasted globally (HLPE 2014, 1)
9. Dessalegn (2011, 18) reports lease rates per hectare per year in 2009 between 1.20 USD and 12 USD.
10. For reflections on customary coping mechanisms and food relief in the past, see Turton (1985).
11. The cases of investors who 'pulled out' need further study. In a panel organized by the Lands of the Future Research Network at the Conference 'Future Africa', African Studies Association in Germany (VAD) Bayreuth University in 2014, several studies mentioned investment failures.
12. 'New research by the consultancy TMP Systems examined 32 cases of tenure conflict in West, East, and Southern Africa, and compared these with 281 cases outside the continent. More than 69 percent of the African conflicts involved a delay in operations and a subsequent loss of money for investors, well above the global average of 56 percent' (Rights and Resources Initiative 2017, 7).
13. Further analysis of these tensions and their relation to development efforts and programmes is needed.
14. 'utopia, n.'. *OED Online*, June 2017, Oxford University Press (http://www.oed.com/view/Entry/220784?redirectedFrom=utopia, accessed August 29, 2017).
15. The rich history of Ethiopia has been described by many outstanding scholars, among them the late Richard Pankhurst whose contribution to Ethiopian historical studies is especially vivid as I am writing this chapter in February 2017, the month of his passing (see Pankhurst 1998).
16. The rectification of historically unjust relationship is laid out in the preamble of the Ethiopian constitution. Here lies the opportunity to

address injustices done to people in the peripheries during the southward expansion of Emperor Menelik in the late nineteenth century (see Lydall 2010).
17. One example is 'no till agriculture' which is widely practiced in Southern Ethiopia. The advantages of 'no till', which can prevent soil from erosion, binds high quantities of CO_2 and can be economically profitable, were rediscovered and praised as technological 'revolutions' for North and South American farmers in the 1980s and 1990s (see also Derpsch 2003; Kassam et al. 2015).
18. I will provide a more detailed discussion on the presence of the Oil Firm 'Tullow' at Arbore in a separate article.
19. I thank the former German Federal Minister for the Environment Klaus Töpfer for discussions and insights into Western environmental debt, due respect and economic realism when it comes to criticizing environmental politics of 'developing' countries.
20. Some Arbore families have started to raise chicken in the past decade.
21. For a meticulous study of the integrated Arbore agricultural and food system, see Miyawaki (1996).
22. Animal health is said to have decreased in the past two decades. A possible relation to resistance to antibiotics has to be verified. The uncontrolled selling of antibiotics for animals by traders has decreased in the past years, since several Arbore who have become veterinary health workers have significantly improved animal health services.
23. This chapter was party inspired by Foucault's interview given anonymously, especially by this passage. 'I can't help but dream about a kind of criticism that would try not to judge but to bring an oeuvre, a book, a sentence, an idea to life; it would light fires, watch the grass grow, listen to the wind, and catch the sea foam in the breeze and scatter it. It would multiply not judgments but signs of existence; it would summon them, drag them from their sleep. Perhaps it would invent them sometimes - all the better. All the better (Foucault 1997, 323)'.

References

Abbink, J. (2011). 'Land to foreigners': Economic, legal, and socio-cultural aspects of new land acquisition schemes in Ethiopia. *Journal of Contemporary African Studies, 29*(4), 513–535.

Abbink, J., Askew, K., Dori, D. F., Gabbert, E., Galaty, J., LaTosky, S. et al. (2014). *Lands of the future. Transforming pastoral lands and livelihoods in Eastern Africa* (MPI Working Papers 154). Halle/Saale: Max Planck Institute for Social Anthropology.

Abebe, M., & Solomon, B. (2013). The need to strengthen land laws in Ethiopia to protect pastoral rights. In A. Catley, J. Lind, & I. Scoones (Eds.), *Pastoralism and development in Africa: Dynamic change at the margins* (pp. 57–70). London and New York: Routledge.

AFSA. (2016). *Agroecology: The bold future of farming in Africa*. Dar es Salaam, Tanzania: Alliance for Food Sovereignty in Africa and Tanzania Organic Agriculture Movement.

Altieri, M. A. (2009). Agroecology, small farms, and food sovereignty. *Monthly Review, 61*(3), 102–111.

Appadurai, A. (2013). *The future as cultural fact: Essays on the global condition*. London: Verso.

Asnake, K., & Fana, G. (2012). Discrepancies between traditional coping mechanisms to climate change and government intervention in South Omo, Ethiopia. In M. G. Berhe & J.-B. Butera (Eds.), *Climate change and pastoralism: Traditional coping mechanisms and conflict in the Horn of Africa* (pp. 123–152). Addis Ababa: Institute for Peace and Security Studies, University for Peace.

Avery, S. (2012). *Lake Turkana and the Lower Omo: Hydrological impacts of major dam and irrigation development* (Report). Oxford: African Studies Centre.

Ayalew, G. (1995). *The Arbore of Southern Ethiopia: A study of inter-ethnic relations, social organization and production practices* (MA thesis, Addis Ababa University, Department of Sociology, Anthropology and Social Administration, Addis Ababa).

Baumann, Z. (2000). *Liquid modernity*. Malden: Polity Press.

Baviskar, A. (Ed.). (2008). *Contested grounds. Essays on nature, culture and power*. Oxford and New York: Oxford University Press.

BMUB und BMEL. (2017). *Nitratbericht 2016*. Bonn: Bundesministerium für Umwelt, Naturschutz, Bau und Reaktorsicherheit, Bundesministerium für Ernährung und Landwirtschaft.

Borras, S. M. Jr., & Franco, J. C. (2012). Global land grabbing and trajectories of agrarian change: A preliminary analysis. *Journal of Agrarian Change, 12*(1), 34–59.

Buffavand, L. (2016). 'The land does not like them': Contesting dispossession in cosmological terms in Mela, South-West Ethiopia. *Journal of Eastern African Studies, 10*(3), 476–493.

Catley, A., Lind, J., & Scoones, I. (Eds.). (2013). *Pastoralism and development in Africa: Dynamic change at the margins*. London and New York: Routledge.

Chivian, E., & Bernstein, A. (Eds.). (2008). *Sustaining life: How human health depends on biodiversity*. Oxford and New York: Oxford University Press.

Coleman-Jensen, A., Rabbitt, M. P., Gregory, C. A., & Singh, A. (2016). *Household food security in the United States in 2015*. Economic Research Service (Economic Research Report Number 215). Washington, DC: United Stated Department of Agriculture.

Cotula, L. (2012). The international political economy of the global land rush: A critical appraisal of trends, scale, geography and drivers. *Journal of Peasant Studies, 39*(3–4), 649–680.

Dereje, F. (2013). 'Centering the periphery'? The federal experience at the margins of the Ethiopian state. *Ethiopian Journal of Federal Studies, 1*(1), 155–192.

Derpsch, R. (2003, January 27–28). Economics of no-till farming. Experiences from Latin America. *Proceedings: No-till on the plains 2003 winter conference* (pp. 43–52). Salina, KS.

Dessalegn, R. (2011). *Land to investors: Large-scale land transfers in Ethiopia* (Discussion Paper). Addis Ababa: Forum for Social Studies.

Dessalegn, R. (2014). Large-scale land investments revisited. In D. Rahmato, M. Ayenew, A. Kefale, & B. Habermann (Eds.), *Reflections on development in Ethiopia: New trends, sustainability and challenges* (pp. 219–245). Addis Ababa: Forum for Social Studies, Friedrich Ebert Stiftung.

Dunlap, R. E. (2013). Climate change skepticism and denial: An introduction. *American Behavioral Scientist, 57*(6), 691–698.

Fana, G. (2016). Land acquisitions, the politics of dispossession, and state-remaking in Gambella, Western Ethiopia. *Africa Spectrum, 51*(1), 5–28.

FAO (Food and Agriculture Organization of the United States). (2014). *Family farmers. Feeding the world, caring for the Earth*. www.fao.org/docrep/019/mj760e/mj760e.pdf. Accessed February 14, 2017.

Foucault, M. (1997). *Ethics: Subjectivity and truth* (Paul Rabinow, Ed.). New York: The New Press.

Fratkin, E. (2014). Ethiopia's pastoralist policies: Development, displacement, and resettlement. *Nomadic Peoples, 18*(1), 94–114.

Gabbert, E. C. (2010). Introduction. In E. Christina & S. Thubauville (Eds.), *To live with others: Essays on cultural neighborhood in Southern Ethiopia* (pp. 13–28). Cologne: Köppe.

Gabbert, E. C. (2014). The global neighbourhood concept. A chance for cooperative development or *festina lente*. In M. G. Berhe (Ed.), *A delicate balance. Land use, minority rights and social stability in the Horn of Africa* (pp. 14–37). Addis Ababa: Institute for Peace and Security Studies, Addis Ababa University.

Galaty, J. G. (2013). Land grabbing in the Eastern African rangelands. In A. Catley, J. Lind, & I. Scoones (Eds.), *Pastoralism and development in Africa: Dynamic change at the margins* (pp. 143–153). London and New York: Routledge.

Gil-Romera, G., Turton, D., & Sevilla-Callejo, M. (2011). Landscape change in the Lower Omo Valley, Southwestern Ethiopia: Burning patterns and woody encroachment in the savanna. *Journal of Eastern African Studies, 5*(1), 108–128.

Giraldo, O. F., & Rosset, P. M. (2017). Agroecology as a territory in dispute: Between institutionality and social movement. *Journal of Peasant Studies*. https://doi.org/10.1080/03066150.2017.1353496. Accessed September 3, 2017.

Girke, F. (2013). *Homeland, boundary, resource: The collision of place-making projects on the Lower Omo River, Ethiopia* (MPI Working Papers 148). Halle/Saale: Max Planck Institute for Social Anthropology.

Gossa, T. W. (1999). *Warfare and fertility. A study of the Hor (Arbore) of Southern Ethiopia* (PhD thesis, School of Economics and Political Science, London).

GRAIN. (2017). *Grabbing the bull by the horns: It's time to cut industrial meat and dairy to save the climate* (Report). Barcelona: GRAIN.

Hallam, E., & Ingold, T. (Eds.). (2007). *Creativity and cultural improvisation* (ASA monographs 44). Oxford and New York: Berg.

Harvey, D. (1996). *Justice, nature and the geography of difference*. Malden and Oxford: Blackwell.

HLPE. (2014). *Food losses and waste in the context of sustainable food systems* (A report by the High Level Panel of Experts on Food Security and Nutrition of the Committee on World Food Security). Rome.

HRW. (2012). *'What will happen if hunger comes?' Abuses against the indigenous peoples of Ethiopia's Lower Omo Valley*. New York: Human Rights Watch.

IFAD. (2013). *Smallholders, food security and the environment*. Rome: International Fund for Agricultural Development.

Ingold, T. (2000). *The perception of the environment. Essays on livelihood, dwelling and skill*. London and New York: Routledge.

Kamski, B. (2016). The Kuraz Sugar Development Project (KSDP) in Ethiopia: Between 'sweet visions' and mounting challenges. *Journal of Eastern African Studies, 10*(3), 568–580.

Kassam, A., Friedrich, T., Derpsch, R., & Kienzle, J. (2015). *Overview of the worldwide spread of conservation agriculture* (Field Actions Science Reports, Vol. 8). http://factsreports.revues.org/3966. Accessed March 1, 2017.

Koohafkan, P., Altieri, M. A., & Gimenez, E. H. (2012). Green agriculture: Foundations for biodiverse, resilient and productive agricultural systems. *International Journal of Agricultural Sustainability, 10*(1), 61–75.

Krätli, S. (2015). *Valuing variability: New perspectives on climate resilient drylands development* (H. de Jode, Ed.). London: International Institute for Environment and Development (IIED).

Krätli, S., & Dyer, C. (2009). *Mobile pastoralists and education: Strategic options* (Education for Nomads Working Paper 1). London: International Institute for Environment and Development (IIED).

Lane, C. R. (Ed.). (1998). *Custodians of the commons: Pastoral land tenure in East and West Africa*. Abingdon and New York: Earthscan.

Lavers, T. (2012a). 'Land grab' as development strategy? The political economy of agricultural investment in Ethiopia. *Journal of Peasant Studies, 39*(1), 105–132.

Lavers, T. (2012b). Patterns of agrarian transformation in Ethiopia: State-mediated commercialization and the 'land grab'. *Journal of Peasant Studies, 39*(3–4), 795–822.

Little, P. D., Smith, K., Cellarius, B. A., Coppock, D. L., & Barrett, C. B. (2001). Avoiding disaster: Diversification and risk management among East African herders. *Development and Change, 32*(3), 401–433.

Lydall, J. (2010). The paternalistic neighbor. A tale of the demise of cherished traditions. In E. Christina & S. Thubauville (Eds.), *To live with others: Essays on cultural neighborhood in Southern Ethiopia* (pp. 314–334). Cologne: R. Köppe.

Magdoff, F., Foster, J. B., & Buttel, F. H. (Eds.). (2000). *Hungry for profit. The agribusiness threat to farmers, food, and the environment*. New York: Monthly Review Press.

Makki, F. (2012). Power and property: Commercialization, enclosures, and the transformation of agrarian relations in Ethiopia. *Journal of Peasant Studies, 39*(1), 81–104.

Makki, F. (2014). Development by dispossession: *Terra nullius* and the social-ecology of new enclosures in Ethiopia. *Rural Sociology, 79*(1), 79–103.

Manzano, P. (2015). Pastoralist ownership of rural transformation: The adequate path to change. *Development, 58*(2–3), 326–332.

Marquard, O. (2003). *Zukunft braucht Herkunft. Philosophische Essays*. Stuttgart: Reclam.

Meckelburg, A. (2014). Large scale land investment in Gambella, Western Ethiopia: The politics and policies of land. In M. G. Berhe (Ed.), *A delicate balance. Land use, minority rights and social stability in the Horn of Africa* (pp. 144–165). Addis Ababa: Institute for Peace and Security Studies, Addis Ababa University.

Miyawaki, Y. (1996). Cultivation strategy and historical change of sorghum varieties in the Hoor of Southwestern Ethiopia. *Essays in Northeast African Studies Senri Ethnological Studies, 43*, 77–120.

MOFED (Ministry of Finance and Economic Development), FDRE. (2010). *Growth and transformation plan (2010/11–2014/15) Vol. I: Main text*, Addis Ababa. http://ethioembassy.org.uk/new/images/About-Ethiopia/GTP%20English%20Vol1.pdf. Accessed December 11, 2016.

National Planning Commission, FDRE. (2016). *Growth and transformation plan II (GTP II) (2015/16–2019/20) Vol. I: Main text*, Addis Ababa. http://dagethiopia.org/new/images/DAG_DOCS/GTP2_English_Translation_Final_June_21_2016.pdf. Accessed March 4, 2017.

Oakland Institute. (2011). *Understanding land investment deals in Africa: Country report: Ethiopia*. Oakland: The Oakland Institute and Stillwater; The Solidarity Movement for a New Ethiopia.

Oakland Institute. (2013). *Omo: Local tribes under threat. A field report from the Omo Valley*. Oakland: Oakland Institute.

Pankhurst, R. (1998). *The Ethiopians. A history*. Malden and Oxford: Blackwell.

Parson, E. A., & Zeckhauser, R. (1995). Cooperation in the unbalanced commons. In K. J. Arrow, R. H. Mnookin, L. Ross, A. Tversky, & R. B. Wilson (Eds.), *Barriers to conflict resolution* (pp. 212–234). New York and London: W. W. Norton.

Rights and Resources Initiative. (2017). *From risk and conflict to peace and prosperity: The urgency of securing community land rights in a turbulent world.* Washington, DC: Rights and Resources Initiative.

Schlee, G. (1989). The orientation of progress. Conflicting aims and strategies of pastoral nomads and development agents in East Africa. A problem survey. In E. Linnebuhr (Ed.), *Transition and continuity of identity in East Africa and beyond—In memoriam David Miller* (pp. 397–450). Bayreuth: E. Breitinger.

Schlee, G. (2013a). Why states still destroy pastoralism and how they can learn that in their own interest they should not. *Nomadic Peoples, 17*(2), 6–19.

Schlee, G. (2013b). Territorializing ethnicity: The imposition of a model of statehood on pastoralists in Northern Kenya and Southern Ethiopia. *Journal of Ethnic and Racial Studies, 36*(5), 857–874.

Scoones, I. (2004). Climate change and the challenge of non-equilibrium thinking. *IDS Bulletin, 35*(3), 114–119.

Scott, J. C. (1998). *Seeing like a state: How certain schemes to improve the human condition have failed.* New Haven and London: Yale University Press.

Strecker, I. (2014). Implications of the international investors code of conduct: The case of the South Ethiopian rift valleys. In M. G. Berhe (Ed.), *A delicate balance. Land use, minority rights and social stability in the Horn of Africa* (pp. 38–63). Addis Ababa: Institute for Peace and Security Studies, Addis Ababa University.

Tewolde, W., & Fana, G. (2014). Social-political and conflict implications of sugar development in Salamago Wereda. Ethiopia. In M. G. Berhe (Ed.), *A delicate balance. Land use, minority rights and social stability in the Horn of Africa* (pp. 117–143). Addis Ababa: Institute for Peace and Security Studies, Addis Ababa University.

Turton, D. (1985). Mursi response to drought: Some lessons for relief and rehabilitation. *African Affairs, 84*(336), 331–346.

Turton, D. (2011). Wilderness, wasteland or home? Three ways of imagining the Lower Omo Valley. *Journal of Eastern African Studies, 5*(1), 158–176.

Turton, D. (2015, August 24–28). *Hydropower and irrigation development in the Omo valley: Development for whom?* Paper presented at the 19th International Conference on Ethiopian Studies. "Ethiopia—Diversity and Interconnections through Space and Time", Warsaw, 2015.

TWN and SOCLA. (2015). *Agroecology: Key concepts, principles and practices.* Penang: Third World Network and Berkeley: Sociedad Científica Latinoamericana de Agroecología.

UN. (2013). *Wake up before it is too late. Make agriculture truly sustainable now for food security in a changing climate* (UNCTAD, Trade and Environment Review Series). New York and Geneva: United Nations.

Wagstaff, Q. A. (2015). *Development, cultural hegemonism and conflict generation in Southwest Ethiopia: Agro-pastoralists in trouble* (Observatoire des Enjeux Politiques et Sécuritaires dans la Corne de l'Afrique. Note 13). Bordeaux: Les Afriques dans le Monde (LAM)/Sciences Po Bordeaux.

White, B., Borras, S. M. Jr., Hall, R. Scoones, I., & Wolford, W. (2012). The new enclosures: Critical perspectives on corporate land deals. *Journal of Peasant Studies, 39*(3–4), 619–647.

Yidnekachew, A. (2015). Policies and practices of consultation with pastoralist communities in Ethiopia: The case of Omo-Kuraz sugar development project. In Y. Aberra & M. Abdulahi (Eds.), *The Intricate road to development. Government development strategies in the pastoral areas of the Horn of Africa* (pp. 274–297). Addis Ababa: Institute for Peace and Security Studies.

Zinsstag, J., Schelling, E., Bonfoh, B., Crump, L., & Krätli, S. (2016). The future of pastoralism: An introduction. *Revue Scientifique et Technique (International Office of Epizootics), 35*(2), 335–355.

Index

A
agri-business 88, 99
agro-pastoralism 18, 20, 139, 143, 150, 167, 208–212, 215, 216, 224, 233, 235, 236, 293, 296
agro-pastoral production system 208
Arbore (people) 299–305, 307, 310
arrowroots 16, 57, 63, 64, 66, 68, 69, 76–79
artisanal mining 138

B
Bamileke (people) 17, 181–183, 185, 189, 190, 201
Bamoum (people) 182, 186, 190
biodiversity 2, 3, 5, 6, 11, 17, 21, 23, 24, 41, 47, 105, 114–116, 118, 119, 129–133, 140–142, 157, 167, 170, 180, 193, 255, 256, 290, 291, 297, 298, 308, 309

conservation 114–116, 118, 129, 131, 132, 256
bridewealth 212, 216, 226, 230

C
cashew 15, 29, 30, 35–41, 44–47, 86
  exports 40
  orchards 30, 35, 40, 42
  production 15, 29, 30, 35, 39, 43, 44
  wine 37, 39, 42
cassava 57, 86, 101, 149, 184, 247, 272
cattle 18, 57, 61, 63, 94, 104, 105, 143, 147–149, 151, 152, 156, 157, 159, 160, 162, 163, 210, 212, 214, 219, 221–224, 228–230, 233–235, 237, 246, 248, 249, 258, 259, 301–303
  raiding 147

rustling 248
trading 219, 224
ceremonies 34, 37, 41, 86, 228
charcoal 127, 129, 133, 212, 219, 220, 225, 245, 249, 251, 259
chemicals 3, 16, 89, 92, 97–100, 103–107, 158, 193, 273, 292
climate change 2, 3, 6, 19, 21, 22, 24, 45, 52, 76, 86, 137, 144, 164, 193, 195, 201, 259, 269–274, 276–284, 290–292, 295, 296, 305
coffee 143, 147, 155, 184, 190
commercialization 12, 15, 16, 207, 209, 214, 229
commoditization 209, 212, 225, 230
conservation and development 54, 244
cotton sector 86, 89, 106
crop failure 210, 217, 220, 223, 228, 229, 302
cultivation 2, 16, 20, 31, 38, 42, 45, 57, 78, 79, 88, 101, 103, 106, 107, 138, 139, 143, 145–150, 152, 156, 157, 161–165, 168, 185, 186, 210, 211, 216–218, 224, 225, 230, 234, 235, 241, 243, 246–250, 252, 254, 258, 293, 296, 300–302
  river bank 143
  shifting 143, 186, 210, 216, 217, 243, 300
cultural ecology 11, 12, 20, 22, 142, 150
culture 11, 12, 17, 20, 140, 166, 189, 216, 270, 274–276, 283, 302, 305
cyclone 13, 245, 250, 251, 253, 254, 257, 258, 261

D
deforestation 18, 179, 180, 195, 235, 236, 241, 243, 251, 254, 257
de-liberalisation 89, 106
dependence 15, 42, 72, 189, 293
'development' 8, 19–21, 43, 47, 138, 160, 170, 306
dikes 31, 32, 38–40, 42, 46, 47
diversification 65, 86, 127
diviner 250
division of labor 32, 37–40
Dizi people (Ethiopia) 159
Dodoma Region 209–212, 230, 232, 235
drought 19, 53, 148, 210, 211, 214, 217, 225, 228, 248, 253, 297, 301, 302

E
ecological adaptability 211
eco-system 6, 140
ecotourism 17, 79, 113–124, 128, 129, 133, 256
  urban-based ecotourism 113, 114, 116, 120, 124
elders 39, 301, 307
entrepreneurship 122, 126, 127
environmental veducation 107, 198
environmental management 7, 59, 60, 141, 163, 193
erosion 9, 18, 30, 52, 62, 78, 157, 158, 180, 198, 236, 251, 310
export economy 43

F
farmer-herder dichotomy 208

floods 31, 157, 236, 245, 250, 253, 258, 273
food export 291
food insecurity 15, 19, 39, 47, 148, 149, 164, 217, 258, 290, 291
food security 39, 40, 44–46, 52, 62, 85, 154–157, 163, 166, 288, 291, 297, 299, 304, 306
forest agrarian systems 241, 243, 245
forest-based livelihoods 244, 245, 259

G

GDP (gross domestic product) 1, 20, 29, 30, 44, 85, 117
Gedeo people (Ethiopia) 167
'global neighbourhood' 287–289
Gogo (people) 208–212, 214, 216, 224, 230, 232–235, 237, 238
government indifference 283

H

headman (headmen) 32, 39, 48
herbicides 16, 88, 89, 91, 92, 95, 97–106
hydrological regulation 78

I

industrialized agriculture 291
industrial processing 36
informal herbicide trade 99
intensification 14, 53, 148, 215, 236, 251
invasive species 52, 105, 162
investors 7, 118–121, 125, 141, 142, 145, 150, 159, 161, 164, 166, 208, 209, 215, 231–233, 236, 287, 288, 292, 294, 309
Iraqw (people) 210, 211, 214, 231–234, 238
irrigation 52, 53, 56, 59–64, 66, 68, 69, 75, 76, 79, 145, 156–158, 186, 246, 250, 260, 293, 300, 301

K

Kwegu (people) 138, 156, 162, 164, 168

L

labor-saving technologies 16
land appropriation 106, 207, 209, 233
land dispute 207
land purchase 225, 226
land use 19, 77, 102, 137, 148, 152, 155, 156, 159, 181, 194, 196, 215, 216, 224, 228, 236, 243, 244, 260, 288, 294, 296, 297, 300, 304–306
leadership 43, 154, 163, 180, 275
lineage elders 30
livestock 15, 16, 18, 34, 58, 59, 61–63, 77, 94, 104–107, 128, 143–145, 147–150, 152, 156, 158, 160, 161, 163, 167, 208–211, 214, 216, 217, 219–223, 229, 234–237, 252, 258, 296–298, 302, 303
local knowledge 3, 4, 6, 55, 139, 143, 166, 290, 292, 306, 307
local resistance 294

M

Maasai (people) 208, 210, 230–233, 235, 236
maize (corn) 18, 57, 61, 67, 69, 86, 87, 89, 94, 100–102, 146, 149, 159–161, 163, 214, 215, 217, 220, 221, 223, 225, 232–237, 241, 246–248, 252–254, 300
mangrove 31, 32, 34, 43, 46, 274
mangrove belt 31
Manjako people (Guinee-Bissau) 29, 31
market 2, 6, 31, 35, 36, 53, 66, 76, 89, 92, 99, 105, 118, 121, 127, 140, 147, 149, 150, 169, 184, 190, 200, 212, 214, 215, 230, 232, 235, 246, 252, 254, 272, 292, 298
marriage 71–75, 148, 302
matrilineal corporate groups 32
mechanization 87, 88, 90
Me'en people (Ethiopia) 145, 147
Mela people (Ethiopia) 141, 158, 169
millet 57, 94, 101, 210, 214, 215, 217
modernization 10, 13, 18, 138, 167, 215, 236, 296
monoculture 15, 40, 53, 215, 236
multi-storey housing 197

N

nature conservation 17, 113, 128, 129, 131–133, 242, 255
Niger delta 269–271, 273, 274
non-farm income 61
Nyangatom people (Ethiopia) 148

O

oil politics 283
oil spillages 273
Omo River 141–143, 148, 156–158, 164, 167, 168, 293
outmigration 45, 105

P

palm oil 30, 34, 35, 159, 170
palm wine 35, 48
pastoralism 138, 143, 145, 160, 166, 208, 210, 211, 214, 233, 235, 236, 249, 303
  transhumant pastoralism 138, 160
performance 279, 280, 282
pesticides 41, 88, 90, 97, 99, 104–106, 158, 292
plantations 10, 23, 57, 103, 138, 143, 145, 152, 154–159, 161–163, 169, 171, 183, 189, 190, 293
play (drama) 19
political ecology 4, 8, 11, 12, 16, 17, 20, 90, 91, 94, 139, 142, 164, 167
pollution 9, 23, 52, 106, 158, 180, 196, 275, 284, 289
population density 17, 183, 185, 188, 212, 213, 224, 237
population growth 2, 3, 7, 9, 13, 14, 18, 19, 21, 23, 93, 103, 144, 149, 151, 153, 170, 213, 242–244, 260
poverty 2, 6, 9, 15, 17, 18, 44, 54, 78, 116, 117, 161, 194, 197, 242, 243, 271, 276, 281, 284, 306

R

rainfall 6, 45, 57, 64, 92, 137, 143, 144, 182, 210, 211, 237, 246, 248, 249, 257, 301
rent-seeking 277
resettlement 17, 158, 160, 164, 170, 190, 212
residential mobility 211
resource utilization 71–75
rice canal 64
rice fields 30–32, 34, 35, 37–40, 42, 43, 46, 48, 77
rice granaries 33, 34
rice harvest 37
risk management 290
ritual leaders 152

S

salinization 52, 157
sorghum 57, 61, 66, 69, 89, 94, 101, 146, 148, 149, 161, 168, 210, 214, 215, 217, 219, 300, 301
spiny forest 18, 245, 247, 255, 260
stakeholder(s) 5, 54, 55, 58, 59, 61, 62, 77, 80, 105, 106, 118, 121, 166, 170, 233, 253, 289, 308
stakeholder analysis 55, 57, 77
state government 40, 44
state policy 7, 13, 44
subsistence farming 53
sugar cane 16, 57, 61, 63, 66, 67, 69, 77, 79, 143, 156–158, 246, 293
sugar cane plantation 156, 157
Suri people (Ethiopia) 17, 138, 139, 162, 168, 171

sustainability 2, 3, 5, 9, 12, 20–22, 54, 79, 91, 140, 141, 146, 154, 164, 184, 291, 298, 306
swidden agriculture 244–247

T

tenure arrangements 248, 257
Toposa (people) 145, 148, 160, 162, 168
tour operators 119–121
tractor(s) 16, 18, 63, 87, 88, 90, 95–97, 99, 102, 214, 215, 217, 232, 234–236
trade-offs 38, 52, 54, 55, 57, 73, 77
traders 39, 40, 47, 66, 143, 149, 150, 310

U

Upland rice 34, 42
urban expansion 53, 189
urban explosion 18

V

villagization 145, 161–163, 212, 214–216

W

Wake up everyone (play) 19, 269–271, 273, 276, 280–283
water pollutants 78
wealth disparity 41
wealth ranking 61
well-being 9, 58, 116, 163, 165, 166, 290, 291, 297, 299, 306

wetland 16, 45, 52–57, 61–64, 66, 68–71, 75, 77–79, 221
wetland products 54, 61, 63, 69, 75, 76
World Bank 6, 7, 9, 15, 41, 44, 45, 138, 147, 194, 195, 201, 202, 215, 242, 244, 250, 251

World Wide Fund for Nature 245

Z

zebu 248, 249, 259

CPSIA information can be obtained
at www.ICGtesting.com
Printed in the USA
LVHW07*1711030618
579426LV00006B/10/P